Selected Titles in This Series

11 **Kunihiko Kodaira, Editor,** Basic analysis: Japanese grade 11, 1996
10 **Kunihiko Kodaira, Editor,** Algebra and geometry: Japanese grade 11, 1996
9 **Kunihiko Kodaira, Editor,** Mathematics 2: Japanese grade 11, 1996
8 **Kunihiko Kodaira, Editor,** Mathematics 1: Japanese grade 10, 1996
7 **Dmitry Fomin, Sergey Genkin, and Ilia Itenberg,** Mathematical circles (Russian experience), 1996
6 **David W. Farmer and Theodore B. Stanford,** Knots and surfaces: A guide to discovering mathematics, 1996
5 **David W. Farmer,** Groups and symmetry: A guide to discovering mathematics, 1996
4 **V. V. Prasolov,** Intuitive topology, 1995
3 **L. E. Sadovskiĭ and A. L. Sadovskiĭ,** Mathematics and sports, 1993
2 **Yu. A. Shashkin,** Fixed points, 1991
1 **V. M. Tikhomirov,** Stories about maxima and minima, 1990

Algebra and Geometry: Japanese Grade 11

昭和57年3月31日文部省検定済
高等学校数学科用

代数・幾何

小平邦彦 編

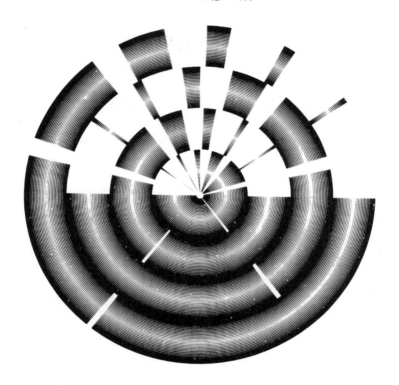

東京書籍株式会社

Mathematical World • Volume 10

Algebra and Geometry
Japanese Grade 11

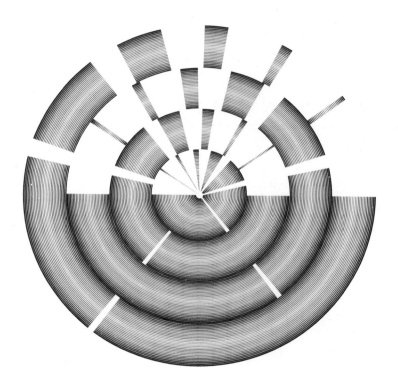

Kunihiko Kodaira, Editor

Hiromi Nagata, Translator
George Fowler, Translation Editor

American Mathematical Society
The University of Chicago School Mathematics Project

The University of Chicago School Mathematics Project
Zalman Usiskin, Director
Izaak Wirszup, Director, Resource Development Component

The translation and publication of this book
were made possible by the generous support of
The Amoco Foundation, Inc.

Translated by Hiromi Nagata
Translation edited by George Fowler

1991 *Mathematics Subject Classification*. Primary 00-01.

Library of Congress Cataloging-in-Publication Data
Daisū, kika. English.
 Algebra and geometry : Japanese grade 11 / edited by Kunihiko Kodaira; Hiromi Nagata, translator; George Fowler, translation editor.
  p. cm. — (Mathematical world, ISSN 1055-9426; v. 10)
 Includes index.
 Summary: A textbook used by upper level secondary school students in Japan, covering plane and solid coordinate geometry, vectors, and matrices.
 ISBN 0-8218-0581-9 (alk. paper)
 1. Algebra. 2. Geometry. [1. Geometry. 2. Algebra.] I. Kodaira, Kunihiko, 1915– . II. Title. III. Series.
QA152.2.D3713 1996
512′.14—dc20   96-20957
                                                                 CIP
                                                                  AC

Copying and reprinting. Individual readers of this publication, and nonprofit libraries acting for them, are permitted to make fair use of the material, such as to copy a chapter for use in teaching or research. Permission is granted to quote brief passages from this publication in reviews, provided the customary acknowledgment of the source is given.

 Republication, systematic copying, or multiple reproduction of any material in this publication (including abstracts) is permitted only under license from the American Mathematical Society. Requests for such permission should be addressed to the Assistant to the Publisher, American Mathematical Society, P. O. Box 6248, Providence, Rhode Island 02940-6248. Requests can also be made by e-mail to `reprint-permission@ams.org`.

Copyright © 1996 by the University of Chicago School Mathematics Project.
All rights reserved. Printed in the United States of America.
Original Japanese edition published in 1984 by Tokyo Shoseki Co., Ltd., Tokyo,
and approved by the Japanese Ministry of Education.
Copyright © 1984 by Tokyo Shoseki Co., Ltd. All rights reserved.
Translated with their permission.
∞ The paper used in this book is acid-free and falls within the guidelines
established to ensure permanence and durability.

10 9 8 7 6 5 4 3 2 1   01 00 99 98 97 96

Textbook Series Preface

The University of Chicago School Mathematics Project

This textbook is part of a series of foreign mathematics texts that have been translated by the Resource Development Component of the University of Chicago School Mathematics Project (UCSMP). Established in 1983 with major funding from the Amoco Foundation, UCSMP has been since that time the nation's largest curriculum development and implementation project in school mathematics. The international focus of its resource component, together with the project's publication experience, makes UCSMP well suited to disseminate these remarkable materials.

The textbooks were originally translated to give U.S. educators and researchers a first-hand look at the content of mathematics instruction in educationally advanced countries. More specifically, they provided input for UCSMP as it developed new instructional strategies, textbooks, and materials of its own; the resource component's translations of over 40 outstanding foreign school mathematics publications, including texts, workbooks, and teacher aids, have been used in UCSMP–related research and experimentation and in the creation of innovative textbooks.

The resource component's translations include the entire mathematics curriculum (grades 1–10) used in the former Soviet Union, standard Japanese texts for grades 7–11, and innovative elementary textbooks from Hungary and Bulgaria.

The content of Japan's compulsory national curriculum for grades 7–11 is made available for the first time in English, thanks in part to the generosity of the Japanese publisher, Tokyo Shoseki Company, Ltd., which provided the copyright permissions.

Japanese Secondary School Mathematics Textbooks

The achievement of Japanese elementary and secondary students gained world prominence largely as a result of their superb performance in the International Mathematics Studies conducted by the International Association for the Evaluation of Educational Achievement. The Second International Mathematics Study surveyed mathematics achievement in 24 countries in 1981–82 and released its findings in 1984. The results are recapitulated in a 1987 national report entitled "The Underachieving Curriculum: Assessing U.S. School Mathematics from an International Perspective" (A National Report on the Second International Mathematics Study, 1987).

Let us take a brief look at the schooling behind much of Japan's economic success. The Japanese school system consists of a six-year primary school, a three-year lower secondary school, and a three-year upper secondary school. The first nine grades are compulsory, and enrollment is now 99.99%. According to 1990 statistics, 95.1% of age-group children are enrolled in upper secondary school, and the dropout rate is 2.2%. In terms of achievement, a typical Japanese student graduates from secondary school with roughly four more years of education than an average American high school graduate. The level of mathematics training achieved by Japanese students can be inferred from the following data:

Japanese Grade 7 Mathematics (New Mathematics 1) explores integers, positive and negative numbers, letters and expressions, equations, functions and proportions, plane figures, and figures in space. Chapter headings in *Japanese Grade 8 Mathematics* include calculating expressions, inequalities, systems of equations, linear functions, parallel lines and congruent figures, parallelograms, similar figures, and organizing data. *Japanese Grade 9 Mathematics* covers square roots, polynomials, quadratic equations, functions, circles, figures and measurement, and probability and statistics. The material in these three grades (lower secondary school) is compulsory for all students.

The textbook *Japanese Grade 10 Mathematics 1* covers material that is compulsory. This course, which is completed by over 97% of all Japanese students, is taught four hours per week and comprises algebra (including quadratic functions, equations, and inequalities), trigonometric functions, and coordinate geometry.

Japanese Grade 11 General Mathematics 2 is intended for the easier of the electives offered in that grade and is taken by about 40% of the students. It covers probability and statistics; vectors; exponential, logarithmic, and trigonometric functions; and differentiation and integration in an informal presentation.

The other 60% of students in grade 11 concurrently take two more extensive courses using the texts *Japanese Grade 11 Algebra and Geometry* and *Japanese Grade 11 Basic Analysis*. The first consists of fuller treatments of plane and solid coordinate geometry, vectors, and matrices. The second includes a more thorough treatment of trigonometry and an informal but quite extensive introduction to differential and integral calculus.

Some 25% of Japanese students continue with mathematics in grade 12. These students take an advanced course using the text *Probability and Statistics* and a second rigorous course with the text *Differential and Integral Calculus*.

One of the authors of these textbooks is Professor Hiroshi Fujita, who spoke at UCSMP's International Conferences on Mathematics Education in 1985, 1988, and 1991. Professor Fujita's paper on Japanese mathematics education appeared in *Developments in School Mathematics Education Around the World*, volume 1 (NCTM, 1987). The current school mathematics reform in Japan is described in the article "The Reform of Mathematics Education at the Upper Secondary School (USS) Level in Japan" by Professors Fujita, Tatsuro Miwa, and Jerry Becker in the proceedings of the Second International Conference, volume 2 of *Developments*.

Acknowledgments

It goes without saying that a publication project of this scope requires the commitment and cooperation of a broad network of institutions and individuals. In acknowledging their contributions, we would like first of all to express our deep appreciation to the Amoco Foundation. Without the Amoco Foundation's generous long-term support of the University of Chicago School Mathematics Project these books might never have been translated for use by the mathematics education community.

We are grateful to UCSMP Director Zalman Usiskin for his help and advice in making these valuable resources available to a wide audience at low cost. Robert Streit, Manager of the Resource Development Component, did an outstanding job coordinating the translation work and collaborating on the editing of most of the manuscripts. George Fowler, Steven R. Young, and Carolyn J. Ayers made a meticulous review of the translations, while Susan Chang and her technical staff at UCSMP handled the text entry and layout with great care and skill. We gratefully acknowledge the dedicated efforts of the translators and editors whose names appear on the title pages of these textbooks.

Izaak Wirszup, Director
UCSMP Resource Development Component

FOREWORD TO THE JAPANESE EDITION

This textbook is intended for students who have completed the study of Mathematics I and go on to study algebra and geometry.

Mathematics was originally linked with science and technology; however, it gradually became independent of science and technology, and present-day mathematicians think freely about virtually everything possible. Therefore, mathematics is said to be a free creation of the human spirit.

On the other hand, mathematics studies mathematical principles which lie behind phenomena in various other fields, as we mentioned in the Foreword to the Mathematics I textbook. Therefore, mathematics is useful because it can be applied to other fields. Mathematics is a very important discipline; it is basic for the study of various other sciences.

Algebra and geometry are important fields; together with analysis, they constitute the foundations of general mathematics. Algebra deals with operations involving expressions, and geometry deals with figures. Although expressions and figures have totally distinct properties, there is nevertheless a close relation between them, as you learned in Mathematics I.

Figures expressed by quadratic equations in terms of plane coordinates are called quadratic curves. Circles are quadratic curves. Parabolas, which you studied in Mathematics I, are also quadratic curves. In Section 1 of Chapter I you will learn the geometric properties of parabolas. Ellipses and hyperbolas are also quadratic curves. In Section 2 of Chapter I you will learn about ellipses and hyperbolas.

In Chapter II we will consider vectors on a plane. In Section 1 you will study the meaning of a vector and operations involving vectors, such as addition, subtraction, and the inner product. In Section 2 you will learn how to apply vectors to figures and other objects.

In Chapter III you will learn about matrices. A matrix is nothing more than an arrangement of numbers or letters in a rectangular or square form, but when we refer to it as a matrix, we regard it as representing a certain quantity, and we define operations involving matrices, such as addition, subtraction, and multiplication. These operations are discussed in Section 1. Matrices with two rows and two columns express linear transformations on a plane. In Section 2 you will study linear transformations on a plane. Matrices play a fundamentally important role in physics. Modern physics regards various quantities as expressed primarily by matrices, rather than by numbers. This is one demonstration that mathematical phenomena underlie natural phenomena.

In Chapter IV you will study figures in space. Section 1 deals with points, straight lines, and planes in space, while Section 2 is devoted to spatial coordinates. In Section 3 you will learn about vectors in space, and in Section 4 you will learn the equations of straight lines and planes in space.

You cannot master mathematics by merely reading books and memorizing; you should think through the material, do calculations, and solve problems by yourselves.

TABLE OF CONTENTS

Chapter 1   **Quadratic Curves**.. 1

       Section 1   Parabolas... 2

             1   Parabolas.. 2
             2   Parabolas and Straight Lines................... 7
             Exercises

       Section 2   Ellipses and Hyperbolas....................... 11

             1   Ellipses 11
             2   Hyperbolas 18
             3   Ellipses, Hyperbolas, and Straight Lines...... 24
             4   Quadratic Curves and Conic Curves 27
             Exercises

       •   Chapter Exercises A and B 30

Chapter 2   **Vectors in the Plane**.................................. 33

       Section 1   Vectors and Operations Involving Them 34

             1   The Meaning of a Vector 34
             2   Addition and Subtraction of Vectors 36
             3   Multiplying a Vector by a Real Number 39
             4   Components of Vectors 43
             5   Inner Product of Vectors 48
             Exercises

       Section 2   Applications of Vectors........................... 57

             1   Position Vectors................................... 57
             2   Straight Lines and Vectors 60
             3   Circles and Vectors 66
             4   Applying Vectors to Figures 68
             5   Force, Velocity, and Vectors................... 70
             Exercises

       •   Chapter Exercises A and B 73

Chapter 3 **Matrices** 75

    Section 1   Matrices 76

        1   The Meaning of a Matrix 76
        2   Addition, Subtraction, and
                Multiplication Involving Matrices 78
        3   Multiplication of Matrices 83
        4   Properties of Multiplication 87
        5   Inverse Matrices 91
        6   Simultaneous Linear Equations 95
        Exercises

    Section 2   Linear Transformations 99

        1   The Meaning of a Linear Transformation 99
        2   The Linearity of Linear Transformations 102
        3   Linear Transformations and Figures 105
        4   The Composite and Inverse of Linear
                Transformations 111
        5   Mapping 114
        Exercises

    •   Chapter Exercises A and B 120

Chapter 4 **Figures in Space** 123

    Section 1   Points, Straight Lines, and Planes in Space .. 124

        1   Straight Lines and Planes in Space 124
        2   Straight Lines Perpendicular to a Plane 127
        Exercises

    Section 2   Coordinates in Space 131

        1   Coordinates in Space 131
        2   The Distance betweeen Two Points,
                and the Equation of a Sphere 134
        Exercises

    Section 3   Vectors in Space 138

        1   Vectors in Space 138
        2   Operations Involving Vectors 139
        3   Components of Vectors 141
        4   The Inner Product of Vectors 146
        Exercises

	Section 4	Equations of Straight Lines and Planes in Space	150
	1	Position Vectors	150
	2	Equation of a Straight Line in Space	153
	3	Equation of a Plane	157
	4	Spheres and Vectors	162
		Exercises	
•		Chapter Exercises A and B	165

- **Answers to Chapter Exercises** 169
- **Index** 171
- **Table of Squares, Square Roots, and Reciprocals** 173
- **Greek Letters** 174

To the Student

(**Example**) This marker designates a concrete example to help you understand the text.

(**Demonstration**) This heading precedes a standard problem for better understanding of the material. Boxes labeled **[Solution]** and **[Proof]** give model answers.

(**Note:**) This marker indicates an explanation to help you understand a particular point.

(**Problem 1**) Problems for rapid mastery of current material and for introducing new material appear in the text with this label.

Exercises At the end of each section problems are provided for practice with the material in that section.

Chapter Exercises At the end of each chapter problems are provided for review of the entire chapter and practical application of the material. A problems involve primarily basic material, while B problems are a bit more advanced.

Reference Refers to topics from other elective courses.

CHAPTER 1

QUADRATIC CURVES

SECTION 1. PARABOLAS
SECTION 2. ELLIPSES AND HYPERBOLAS

A figure in the plane represented by a linear equation in x and y is a straight line. What figures are represented by quadratic equations? In this chapter you will learn the fundamental properties of "quadratic curves."

When you cut a cone, a quadratic curve appears as a section. Therefore, quadratic curves are also called "conic curves," and they have been studied since ancient Greece. In particular, the work of Apollonius (260?-200? B.C.) was precise and renowned. He classified conic sections cut by a plane into three categories, and called them ellipse (literally, "leaving out"), parabola ("comparison"), and hyperbola ("excess"); we have retained these terms to refer to the curves created by these sections.

Besides conic sections, quadratic curves occur as the orbits of the planets and in a number of familiar natural phenomena. Thus, they are indispensable to descriptions of the physical world.

PARABOLAS

 Parabolas

You already learned in Mathematics I that the graph of the quadratic function

$$y = ax^2 + bx + c$$

is a parabola. In this section let's again consider the geometric character of this curve.

A circle, for example, is the locus of points whose distance from a fixed point is a constant. If we consider a parabola as a locus of points, then we can give the following definition of a parabola:

A **parabola** is the locus of points whose distance from a fixed point F is equal to their distance from a fixed line l that does not pass through F. Point F is called the **focus** of the parabola, and line l is the **directrix**.

Next, let's find the equation of a parabola from the above definition.

Draw a line FH from the focus F perpendicular to the directrix l. Then define coordinate axes by taking the midpoint O of FH as the origin, taking the perpendicular bisector of line segment FH as the x-axis, and taking the line FH as the y-axis. For a non-zero p,

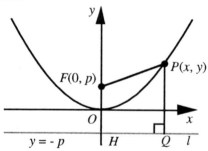

the coordinates of F are $(0, p)$, and

the equation of l is $y = -p$.

Draw line PQ perpendicular to l from point $P(x, y)$ on the locus, and then

$$PQ = |y + p|, \quad PF = \sqrt{x^2 + (y - p)^2}.$$

Therefore, the condition of the locus $PQ = PF$ can be rewritten as

$$|y + p| = \sqrt{x^2 + (y - p)^2}.$$

Squaring both sides of this equation, we obtain

$$(y + p)^2 = x^2 + (y - p)^2.$$

Rearranging this equation,

$$4py = x^2 \quad \text{or} \quad y = \frac{1}{4p}x^2. \tag{1}$$

This is the equation of the parabola.

The origin O is called the vertex of the parabola, and the y-axis is the axis.

The Equation of a Parabola

The equation of a parabola in which the focus is $(0, p)$ and the directrix is $y = -p$ is

$$4py = x^2 \quad \text{or} \quad y = \frac{1}{4p}x^2.$$

Problem 1 What will happen to the parabola above if $p < 0$?

The quadratic function $y = ax^2$ can be rewritten in the form of (1) by setting $a = \frac{1}{4p}$ or $p = \frac{1}{4a}$. Therefore, the graph of $y = ax^2$ is a parabola in which the focus is the point $(0, \frac{1}{4a})$ and the directrix is the line $y = -\frac{1}{4a}$.

Problem 2 Find the equations of the following parabolas:

(1) focus $(0, 2)$, directrix $y = -2$

(2) focus $(0, -\frac{1}{2})$, directrix $y = \frac{1}{2}$

(3) focus $(0, \frac{1}{4})$, vertex $(0, 0)$

1 QUADRATIC CURVES

Problem 3   Find the coordinates of the focus and the equation of the directrix for the following parabolas:

(1) $y = \dfrac{1}{4}x^2$ (2) $y = -2x^2$ (3) $x^2 = 12y$

The curve created by reflecting parabola (1) above with respect to the line $y = x$ is the figure represented by the equation

$$y^2 = 4px.$$

This curve is a parabola with

the point $(p, 0)$ as its focus and

the line $x = -p$ as its directrix.

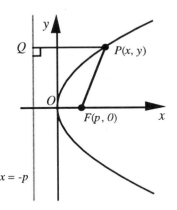

Problem 4   Find the equations of the following parabolas:

(1) focus $(1, 0)$, directrix $x = -1$

(2) focus $(-3, 0)$, directrix $x = 3$

Problem 5   Find the coordinates of the focus and the equation of the directrix for the following parabolas:

(1) $y^2 = x$ (2) $y^2 = -8x$

Problem 6   Sketch the regions represented by the following inequalities:

(1) $y < 4x^2$ (2) $y^2 \leq 4x$ (3) $y^2 > -3x$

Translating Parabolas

As you learned in Mathematics I, the equation of the figure created by translating the curve represented by the equation $f(x, y) = 0$ by m units along the x-axis and n units along the y-axis is given by

$$f(x - m, y - n) = 0.$$

Example The equation of the parabola created by translating the parabola $4py = x^2$, in which the focus is $(0, p)$ and the directrix is $y = -p$, by m units along the x-axis and n units along the y-axis is

$$4p(y - n) = (x - m)^2.$$

The focus of this new parabola is $(m, p + n)$, and the directrix is $y = -p + n$.

Problem 7 Find the equations of the parabolas created by translating the following parabolas by 2 units along the x-axis and -3 units along the y-axis. Find the focus and the directrix.

(1) $2y = -x^2$ (2) $y^2 = 5x$

Demonstration Show that the figures represented by the following equations are parabolas and draw rough sketches of them. Then find the coordinates of the focus and the equation of the directrix.

(1) $y = x^2 - 2x + 2$ (2) $y^2 = 4x + 8$

[Solution] (1) $y - 1 = (x - 1)^2$

This equation represents the parabola created by translating the parabola $y = x^2$ by 1 unit along the x- and y-axes.

focus $(1, \frac{5}{4})$

directrix $y = \frac{3}{4}$

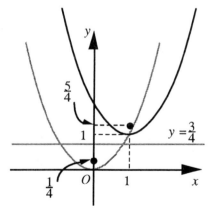

(2) $y^2 = 4(x + 2)$

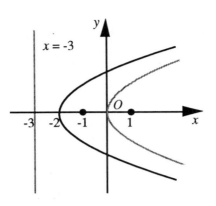

This equation represents the parabola created by translating the parabola $y^2 = 4x$ by -2 units along the x-axis.

focus (-1, 0)

directrix $x = -3$

Problem 8 Sketch the parabolas represented by the following equations. Find the coordinates of the focus and the equation of the directrix.

(1) $4y = -x^2 + 4x + 8$

(2) $y^2 - 6y = x$

Parabolas and Straight Lines

Demonstration 1   How will the number of common points shared by the parabola $y^2 = 4x$ and the straight line $y = x + k$ change according to the value of k?

[Solution]

$$y^2 = 4x \qquad (1)$$

$$y = x + k \qquad (2)$$

The real solutions of the quadratic equation

$$y^2 - 4y + 4k = 0 \qquad (3)$$

created by eliminating x from (1) and (2) are the y coordinates of the common points.

Therefore, the number of common points is equal to the number of real solutions to quadratic equation (3).

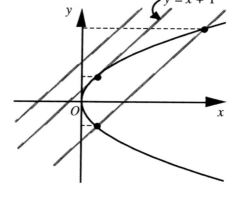

If we take D as the discriminant of equation (3), then

$$\frac{D}{4} = 4 - 4k = 4(1 - k)$$

(i) For $D > 0$ or $k < 1$, they have two common points.

(ii) For $D = 0$ or $k = 1$, they have one common point.

(iii) For $D < 0$ or $k > 1$, they have no common point.

Answer: For $k < 1$, 2 points; for $k = 1$, 1 point; for $k > 1$, no points.

In Demonstration 1, if there is one common point, the graphs of (1) and (2) are said to be **tangent**, and the common point is the **tangent point**.

Problem 1 Find the coordinates of the tangent point in Demonstration 1.

Problem 2 How will the number of common points shared by the parabola $y^2 = -8x$ and the straight line $y = 2x + k$ change according to the value of k?

Demonstration 2 Prove that if the parabola $y^2 = 4x$ and the straight line $y = x + k$ have two common points P and Q, the midpoint R of PQ lies on the straight line $y = 2$ in the part specified by $x > 1$.

[Proof] As we showed in solving Demonstration 1, these figures have two common points P and Q for $k < 1$. Take y_1 and y_2 as y coordinates of P and Q. Since y_1 and y_2 are the solutions of the quadratic equation

$$y^2 - 4y + 4k = 0,$$

the relation between coefficients and solutions gives us

$$y_1 + y_2 = 4.$$

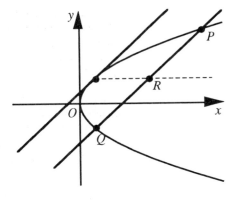

Therefore, if we take (x, y) as the coordinates of R, then

$$y = \frac{y_1 + y_2}{2} = 2$$

$$x = y - k = 2 - k.$$

For $k < 1$, $x = 2 - k > 1$.

Thus, R lies on the line $y = 2$ in the part specified by $x > 1$.

Problem 3 The parabola $y^2 = -3x$ and the straight line $y = \frac{1}{2}x + k$ have two common points P and Q. On what figure does the midpoint R of line segment PQ lie?

Exercises

1. Find the equations of the following parabolas:

    (1) focus $(0, -4)$, directrix $y = 4$

    (2) focus $(2, 0)$, directrix $x = -2$

2. Sketch the following parabolas:

    (1) $y^2 = -2x + 6$ (2) $y^2 - 2y = 4x + 3$

3. Sketch the regions defined by the following inequalities.

    (1) $\begin{cases} y^2 \leq 4x \\ y \geq 2x \end{cases}$ (2) $-4y^2 < x < 2$

4. Given

    the parabola $y^2 = 4x$ (1)

    the straight line $y = \frac{1}{2}x + k$ (2)

    (1) Find the coordinates of the focus F of parabola (1).

    (2) Find the value of k and the coordinates of the tangent point P if (1) and (2) are tangent.

    (3) In the case of (2) take Q as the point at which line (2) intersects the x-axis. Show that $PF = QF$.

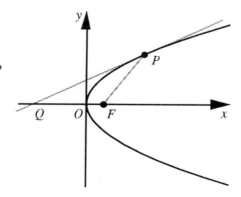

5. How will the number of common points shared by the parabola $y^2 = 2x$ and the straight line $y = mx + 1$ change according to the value of m?

6. Take F as the focus of the parabola $y^2 = 4px$, and take P as an arbitrary point on this parabola. On what figure does the midpoint R of line segment FP lie?

ELLIPSES AND HYPERBOLAS

 ### Ellipses

An **ellipse** is the locus of a point such that the sum of its distances from two fixed points F and F' is a constant. Points F and F' are each said to be a **focus** of the ellipse.

We can draw an ellipse by moving a pencil along the outermost path permitted by a string of fixed length with its ends fixed at F and F', as in the figure to the right.

An ellipse is symmetric with respect to the straight line connecting the two foci or the perpendicular bisector of the line segment connecting the two foci. Therefore, it is also symmetric with respect to the midpoint O of the line segment connecting the two foci. This point O is said to be the **center** of this ellipse.

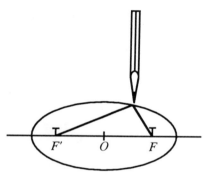

A circle can be considered as a special case of an ellipse where the foci coincide.

Next, let us find the equation of an ellipse by taking the center O as the origin, the line FF' connecting the foci as the x-axis, and the perpendicular bisector of line segment FF' as the y-axis.

Take $(c, 0)$ and $(-c, 0)$ as the coordinates of F and F' and $2a$ as the given fixed length. Provided that $c > 0$ and $a > c$, we can draw the figure. Take point $P(x, y)$ on the locus. Since

$$PF = \sqrt{(x - c)^2 + y^2}, \; PF' = \sqrt{(x + c)^2 + y^2},$$

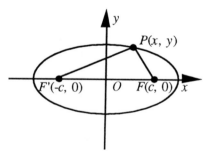

the condition of the locus $PF + PF' = 2a$ can be written as

$$\sqrt{(x - c)^2 + y^2} + \sqrt{(x + c)^2 + y^2} = 2a. \tag{1}$$

Multiplying both sides of (1) by $\sqrt{(x-c)^2+y^2} - \sqrt{(x+c)^2+y^2}$, we obtain

$$\{(x-c)^2+y^2\} - \{(x+c)^2+y^2\} = 2a\,(\sqrt{(x-c)^2+y^2} - \sqrt{(x+c)^2+y^2}).$$

The left side of this equation is equal to $-4cx$, and therefore

$$\sqrt{(x-c)^2+y^2} - \sqrt{(x+c)^2+y^2} = -\frac{2cx}{a}. \tag{2}$$

Adding both sides of (1) to the corresponding sides of (2), and then dividing them by 2, we obtain

$$\sqrt{(x-c)^2+y^2} = a - \frac{c}{a}x.$$

Squaring both sides and rearranging them, we get

$$\frac{a^2-c^2}{a^2}x^2 + y^2 = a^2 - c^2.$$

Therefore,

$$\frac{x^2}{a^2} + \frac{y^2}{a^2-c^2} = 1.$$

Since $a > c > 0$, we can take $a^2 - c^2 = b^2$ and $b > 0$, and then

$$\frac{x^2}{a^2} + \frac{y^2}{b^2} = 1. \tag{3}$$

The Equation of an Ellipse

The equation of an ellipse, the locus of a point such that the sum of its distances from the two foci $(c, 0)$ and $(-c, 0)$ is $2a$, is

$$\frac{x^2}{a^2} + \frac{y^2}{b^2} = 1$$

with $a > c > 0$ and $b = \sqrt{a^2 - c^2}$.

Equation (3) is referred to as the **standard form** for the equation of an ellipse.

Example The locus of a point such that the sum of its distances from two points $F(4, 0)$ and $F'(-4, 0)$ is 10, since

$$a = \frac{10}{2} = 5, \quad b^2 = a^2 - c^2 = 25 - 16 = 9$$

is the ellipse $\frac{x^2}{25} + \frac{y^2}{9} = 1$.

Problem 1 Find the equation of a locus of a point such that the sum of its distances from the two points $(3, 0)$ and $(-3, 0)$ is 10.

In equation (3) above,

if we take $y = 0$, then $x = \pm a$,

and if we take $x = 0$, then $y = \pm b$.

Therefore, ellipse (3) intersects the x-axis at

$A(a, 0)$ and $A'(-a, 0)$

and it intersects the y-axis at

$B(0, b)$ and $B'(0, -b)$.

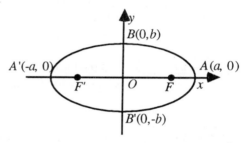

These four points are called the **vertices** of the ellipse.

In the above ellipse, since $a^2 - c^2 = b^2$, we know that $a > b$. In this case, AA' and BB' are referred to as the **major axis** and the **minor axis** of ellipse (3).

Since $c = \sqrt{a^2 - b^2}$, the coordinates of the foci F and F' of ellipse (3) are

$$F(\sqrt{a^2 - b^2}, 0), \quad F'(-\sqrt{a^2 - b^2}, 0).$$

In the equation

$$\frac{x^2}{a^2} + \frac{y^2}{b^2} = 1,$$

when $b > a > 0$, we can see that this equation represents an ellipse in which the foci are two points on the y-axis

$$F(0, \sqrt{b^2 - a^2}),$$

$$F'(0, -\sqrt{b^2 - a^2}),$$

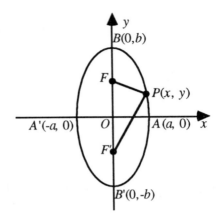

and we have interchanged the roles of a and b and the x- and y-axes. In this case, the sum of the distances from the two foci is $2b$, and AA' and BB' in the figure are the minor and major axes, respectively.

Problem 2 Find the locus of points such that the sum of their distances from the two points $(0, 3)$ and $(0, -3)$ is 10.

Problem 3 Find the vertices and foci of the following ellipses, and draw rough sketches. Then state the length of the major and minor axes.

(1) $\dfrac{x^2}{25} + \dfrac{y^2}{16} = 1$ (2) $x^2 + \dfrac{y^2}{3^2} = 1$

(3) $4x^2 + 9y^2 = 36$ (4) $\dfrac{x^2}{3} + \dfrac{y^2}{4} = 1$

Circles and Ellipses

If $a > b > 0$, then the circle which has the major axis of the ellipse

$$\frac{x^2}{a^2} + \frac{y^2}{b^2} = 1 \qquad (1)$$

as its diameter is

$$x^2 + y^2 = a^2. \qquad (2)$$

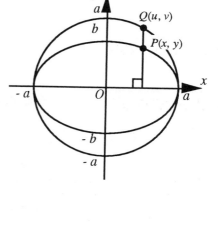

Take $Q(u, v)$ as a point moving along the circumference of circle (2), and $P(x, y)$ as a point whose y–coordinate is $\frac{b}{a}$ times the y-coordinate of Q. Then

$$u^2 + v^2 = a^2 \qquad (3)$$

$$x = u, \quad y = \frac{b}{a}v. \qquad (4)$$

From (4), $\quad u = x, \quad v = \frac{a}{b}y.$

Substituting these values into (3), we obtain

$$x^2 + \left(\frac{a}{b}y\right)^2 = a^2 \quad \text{or} \quad \frac{x^2}{a^2} + \frac{y^2}{b^2} = 1.$$

Thus, point P lies on ellipse (1).

Therefore, ellipse (1) can be considered as defining circle (2) at a ratio of $\frac{b}{a}$ along the y-axis.

Problem 4 In the above case, if we want to find the coordinates of point P from point Q, we can find the intersection R of OQ and the circle $x^2 + y^2 = b^2$, and then find the point at which the perpendicular line from Q to the x-axis intersects the line from R parallel to the x-axis. Demonstrate this method.

1 QUADRATIC CURVES

Problem 5 Sketch the region represented by the following inequalities:

(1) $\dfrac{x^2}{3^2} + \dfrac{y^2}{2^2} < 1$ (2) $x^2 + \dfrac{y^2}{4} \geq 1$

Translating Ellipses

Demonstration 1   What figure does the following equation represent?

$$\dfrac{(x-2)^2}{4} + (y-1)^2 = 1$$

[Solution] The figure represented by this equation is the figure created by translating the ellipse

$$\dfrac{x^2}{4} + y^2 = 1$$

by 2 units along the x-axis and 1 unit along the y-axis. The figure to the right is a sketch of these ellipses.

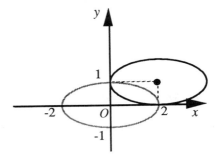

Problem 6 Give the coordinates of the center, the vertices, and the foci of the ellipse in Demonstration 1.

Demonstration 2 Show that the figure represented by the following equation is an ellipse, and sketch it.

$$9x^2 + 4y^2 + 18x - 27 = 0$$

[Solution] Transforming the given equation, we obtain

$$9(x + 1)^2 + 4y^2 = 36$$

$$\frac{(x + 1)^2}{4} + \frac{y^2}{9} = 1.$$

Therefore, the figure represented by this equation is the ellipse created by translating the ellipse $\frac{x^2}{4} + \frac{y^2}{9} = 1$ by -1 unit along the x-axis. A sketch of these ellipses is given to the right.

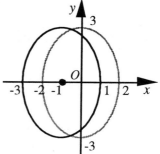

Problem 7 Sketch the figures represented by the following equations:

(1) $(x - 1)^2 + \frac{(y + 1)^2}{2} = 1$

(2) $4x^2 + 9y^2 = 24x$

Hyperbolas

A **hyperbola** is the locus of a point such that the difference of its distances from two fixed points F and F' is a constant, and F and F' are called the **foci**.

Taking $2a$ as the constant, we can draw a hyperbola by the following method. Fix one end of a ruler of length l at F' and allow it to rotate. Fix a string of length $l - 2a$ at the other end of the ruler and at F. Now move a pencil along the outermost path permitted by the string such that the endpoint P of the pencil remains in contact with both the string and the ruler. We obtain curves such as the ones illustrated to the right.

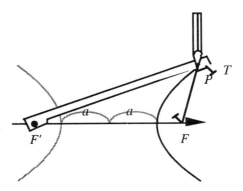

A hyperbola is also symmetric with respect to the straight line connecting the two foci and the perpendicular bisector of the line segment connecting the two foci. Therefore, it is also symmetric with respect to the midpoint O of the line segment connecting the two foci. This point O is referred to as the **center** of the hyperbola.

Just as we did for the case of an ellipse, let us define coordinate axes by taking the center O as the origin, the line FF' connecting the two foci as the x-axis, and the perpendicular bisector of line segment FF' as the y-axis. Now we can find the equation of the hyperbola.

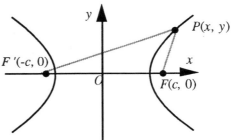

Take $(c, 0)$ and $(-c, 0)$ as the coordinates of F and F', take $2a$ as the fixed length, and let $P(x, y)$ be a point on the locus. Then the condition of the locus $PF' - PF = \pm 2a$ can be expressed as

$$\sqrt{(x + c)^2 + y^2} - \sqrt{(x - c)^2 + y^2} = \pm 2a. \qquad (1)$$

In this case, so that we can actually draw the figure, we must assume that $c > a$.

Just as with an ellipse, by multiplying both sides of (1) by $\sqrt{(x + c)^2 + y^2} + \sqrt{(x - c)^2 + y^2}$ we obtain

$$\sqrt{(x + c)^2 + y^2} + \sqrt{(x - c)^2 + y^2} = \pm \frac{2cx}{a}. \qquad (2)$$

From (1) and (2),

$$\sqrt{(x+c)^2 + y^2} = \pm\left(a + \frac{c}{a}x\right).$$

Squaring both sides and rearranging the terms, we obtain

$$\frac{c^2 - a^2}{a^2}x^2 - y^2 = c^2 - a^2.$$

Therefore,

$$\frac{x^2}{a^2} - \frac{y^2}{c^2 - a^2} = 1.$$

Taking $c^2 - a^2 = b^2$, $b > 0$,

$$\frac{x^2}{a^2} - \frac{y^2}{b^2} = 1. \qquad (3)$$

The Equation of a Hyperbola

The equation of a hyperbola, the locus of points the difference of whose distances from two foci $(c, 0)$ and $(-c, 0)$ is $2a$, is

$$\frac{x^2}{a^2} - \frac{y^2}{b^2} = 1$$

with $c > a > 0$ and $b = \sqrt{c^2 - a^2}$.

The equation in (3) is said to be the **standard form** of the equation of a hyperbola.

As in the figure on the preceding page, hyperbola (3) consists of two parts. The part associated with focus F is the locus of point P such that $PF' - PF = 2a$, and the part associated with focus F' is the locus of point P such that $PF - PF' = 2a$.

Problem 1 Find the equation of the locus of a point such that the difference of its distances from the two points $(3, 0)$ and $(-3, 0)$ is 4.

Hyperbola (3) intersects the x-axis at two points

$$A(a, 0) \text{ and } A'(-a, 0)$$

and it does not intersect the y-axis. Points A and A' are called the **vertices** of hyperbola (3).

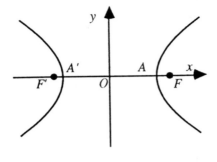

The foci of this hyperbola are

$$F(\sqrt{a^2 + b^2}, 0)$$

$$F'(-\sqrt{a^2 + b^2}, 0).$$

If we consider that we have interchanged the roles of a and b, and the x- and y-axes, then we know that the equation

$$\frac{y^2}{b^2} - \frac{x^2}{a^2} = 1 \text{ or } \frac{x^2}{a^2} - \frac{y^2}{b^2} = -1$$

expresses a hyperbola in which the foci are points on the y-axis

$$(0, \sqrt{a^2 + b^2}), (0, -\sqrt{a^2 + b^2}).$$

The vertices of this hyperbola are

$$B(0, b), B'(0, -b).$$

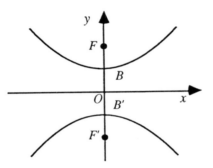

The difference of the distances from the two foci to a point on the hyperbola is $2b$.

Problem 2 Find the coordinates of the foci and vertices of the following hyperbolas:

(1) $\dfrac{x^2}{4} - \dfrac{y^2}{9} = 1$ (2) $\dfrac{x^2}{3^2} - \dfrac{y^2}{4^2} = -1$

Asymptotes

Let's examine the figure of hyperbola (3) in a little more detail.

Solving (3) with respect to y, we obtain

$$y = \pm \frac{b}{a}\sqrt{x^2 - a^2}.$$

The quantity under the radical cannot be negative, so $x \geq a$ or $x \leq -a$.

In order to examine the graph of a hyperbola, it is sufficient to consider only the part for $x \geq a$ and $y \geq 0$

$$y = \frac{b}{a}\sqrt{x^2 - a^2} \quad (x \geq a).$$

In this part of the hyperbola for $x = a$, $y = 0$, as x increases from a, y also increases; and as x increases without limit, y also increases without limit. Since

$$\frac{b}{a}\sqrt{x^2 - a^2} < \frac{b}{a}x, \quad (4)$$

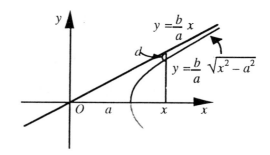

the hyperbola lies below the straight line $y = \frac{b}{a}x$. Take d as the difference of the left and right sides.

$$d = \frac{b}{a}(x - \sqrt{x^2 - a^2}) = \frac{ab}{x + \sqrt{x^2 - a^2}}$$

As x increases without limit, the denominator increases without limit, and therefore d approaches infinitely close to 0. Thus, a point (x, y) on the hyperbola approaches the line $y = \frac{b}{a}x$ as x increases without limit.

Example In the hyperbola $\frac{x^2}{4} - y^2 = 1$, $a = 2$ and $b = 1$. The part of the hyperbola which lies in the first quadrant is

$$y = \frac{1}{2}\sqrt{x^2 - 4}$$

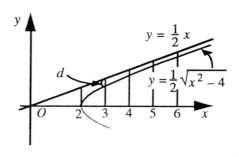

Therefore,

$$d = \frac{1}{2}(x - \sqrt{x^2 - 4}) = \frac{2}{x + \sqrt{x^2 - 4}}.$$

The values of d on the interval $x \geq 2$ are given in the table below. From this table, we can see that the value of d gradually approaches 0 as x increases.

x	d	x	d	x	d	x	d
2	1.0000	5	0.2087	8	0.1270	50	0.0200
3	0.3820	6	0.1716	9	0.1125	100	0.0100
4	0.2679	7	0.1459	10	0.1010	500	0.0020

Analogously, the point (x, y) on the hyperbola $\frac{x^2}{a^2} - \frac{y^2}{b^2} = 1$ also approaches the line $y = \frac{b}{a}x$ in the third quadrant, and it approaches the line $y = -\frac{b}{a}x$ in the second and the fourth quadrants as $|x|$ increases without limit.

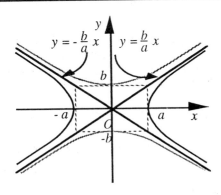

These lines

$$y = \frac{b}{a}x, \quad y = -\frac{b}{a}x$$

are referred to as the **asymptotes** of the hyperbola $\frac{x^2}{a^2} - \frac{y^2}{b^2} = 1$.

The asymptotes of the hyperbola $\frac{x^2}{a^2} - \frac{y^2}{b^2} = -1$ are the same as above.

Problem 3 Find the vertices, asymptotes, and foci of the following hyperbolas. Then sketch them.

(1) $\dfrac{x^2}{16} - \dfrac{y^2}{9} = 1$ (2) $9x^2 - 4y^2 = 1$

(3) $x^2 - y^2 = 2$ (4) $4x^2 - 25y^2 = -100$

(5) $\dfrac{y^2}{9} - \dfrac{x^2}{4} = 1$

The two asymptotes of the hyperbola $\dfrac{x^2}{a^2} - \dfrac{y^2}{a^2} = 1$ or

$$x^2 - y^2 = a^2$$

are $y = x$ and $y = -x$, and they intersect at a right angle. A hyperbola whose asymptotes intersect at a right angle is called a **rectangular hyperbola**.

Problem 4 Find the equations of the rectangular hyperbolas in which the foci are given by the following pairs of points:

(1) (3, 0), (-3, 0)   (2) (0, 2), (0, -2)

Translating Hyperbolas

Demonstration Show that the figure represented by the following equation is a hyperbola and sketch it.

$$x^2 - 2y^2 - 4y = 10$$

[Solution] Transforming the equation, we obtain

$$x^2 - 2(y + 1)^2 = 8$$

$$\dfrac{x^2}{8} - \dfrac{(y + 1)^2}{4} = 1.$$

Therefore, the figure represented by this equation is created by translating the hyperbola $\dfrac{x^2}{8} - \dfrac{y^2}{4} = 1$ by -1 unit along the y-axis.

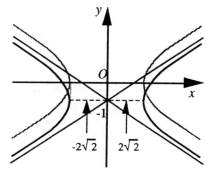

vertices: $(\pm 2\sqrt{2}, -1)$

asymptotes: $y = \pm \dfrac{1}{\sqrt{2}} x - 1$

A sketch of this hyperbola is shown above to the right.

1 QUADRATIC CURVES

Problem 5 Find the coordinates of the foci of the hyperbola in the above Demonstration.

Problem 6 Sketch the figures represented by the following equations:

(1) $\dfrac{(x+1)^2}{16} - \dfrac{(y-2)^2}{9} = 1$ (2) $x^2 - y^2 + 4x + 8 = 0$

  Ellipses, Hyperbolas, and Straight Lines

Demonstration How will the number of common points shared by the ellipse $\dfrac{x^2}{4} + y^2 = 1$ and the straight line $y = \dfrac{1}{2}x + k$ change according to the value of k?

[Solution]

$$\dfrac{x^2}{4} + y^2 = 1 \qquad (1)$$

$$y = \dfrac{1}{2}x + k \qquad (2)$$

The real solutions of the quadratic equation

$$x^2 + 2kx + 2k^2 - 2 = 0 \qquad (3)$$

derived by eliminating y from (1) and (2) are the x-coordinates of the common points. Therefore, the number of common points is equal to the number of real solutions to the quadratic equation (3).

Take D as the discriminant of equation (3).

$$\dfrac{D}{4} = k^2 - (2k^2 - 2)$$

$$= -k^2 + 2$$

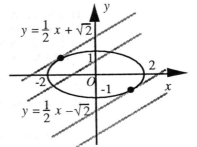

(i) For $D > 0$ or $-\sqrt{2} < k < \sqrt{2}$, there are two common points.

(ii) For $D = 0$ or $k = \pm\sqrt{2}$, there is one common point.

(iii) For $D < 0$ or $k < -\sqrt{2}$ or $\sqrt{2} < k$, there are no common points.

Answer: For $-\sqrt{2} < k < \sqrt{2}$, 2 points;

for $k = \pm\sqrt{2}$, 1 point;

for $k < -\sqrt{2}$ or $\sqrt{2} < k$, no points.

In the special case when the graphs of (1) and (2) have one common point, they are said to be **tangent**, and that common point is called the **tangent point**.

Problem 1 How will the number of common points shared by the ellipse $\dfrac{x^2}{4} + \dfrac{y^2}{2} = 1$ and the straight line $y = x + k$ change according to the value of k?

Problem 2 How will the number of common points shared by the hyperbola $\dfrac{x^2}{4} - y^2 = 1$ and the straight line $y = x + k$ change according to the value of k?

Demonstration 2 Prove that if the ellipse $\dfrac{x^2}{4} + y^2 = 1$ and the straight line $y = \dfrac{1}{2}x + k$ have two intersections P and Q, the midpoint R of line segment PQ lies on the line $y = -\dfrac{1}{2}x$, in the part corresponding to $-\sqrt{2} < x < \sqrt{2}$.

[**Proof**] These figures have two common points P and Q for $-\sqrt{2} < k < \sqrt{2}$, just as in Demonstration 1. Take x_1 and x_2 as the x-coordinates of P and Q, and then x_1 and x_2 are the solutions of the quadratic equation

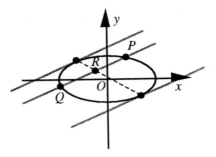

$$x^2 + 2kx + 2k^2 - 2 = 0.$$

Therefore, from the relation between the coefficient and the solutions,

$$x_1 + x_2 = -2k.$$

Thus, if we take (x, y) as the coordinates of R, the midpoint of PQ, then

$$x = \frac{x_1 + x_2}{2} = -k, \quad (1)$$

$$y = \frac{x}{2} + k = \frac{k}{2}. \quad (2)$$

Eliminating k from (1) and (2), we obtain

$$y = -\frac{1}{2}x.$$

Since $-\sqrt{2} < k < \sqrt{2}$,

$$-\sqrt{2} < x < \sqrt{2}.$$

Therefore, R lies on line $y = -\frac{1}{2}x$, in the part corresponding to $-\sqrt{2} < x < \sqrt{2}$.

Problem 3 On what figure does the midpoint R of line segment PQ lie, if the ellipse $4x^2 + y^2 = 4$ and the straight line $y = -x + k$ have two common points P and Q?

Quadratic Curves and Conic Curves

The circle, which we studied in Mathematics I, the parabola, the ellipse, and the hyperbola which we have been studying here, are represented by quadratic equations. Therefore, these curves are referred to as **quadratic curves**.

Moreover, these curves are known to appear as sections of cones cut by a plane. Therefore, these curves are also called **conic curves**.

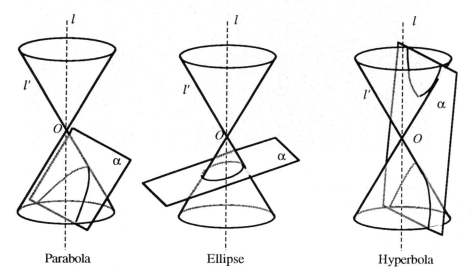

Parabola　　　　　Ellipse　　　　　Hyperbola

If a straight line l' which intersects line l at a single point O is revolved about l in space, the surface that l' describes is called a cone. Line l is referred to as the **axis**, point O is the **vertex**, and line l' is the **generatrix**.

A cone is divided into two parts by the vertex. When a cone is cut by a plane α which does not pass through the vertex, if α is parallel to one of the generatrices, α intersects one part of the cone, and the section thus created is a parabola.

If α is not parallel to any of the generatrices, and it intersects only one part of the cone, the section is an ellipse. And if α intersects both parts of the cone, the section is a hyperbola.

Quadratic Curves and Natural Phenomena

A curve described by a ball thrown upward and forward into the air is a parabola, if we neglect air resistance.

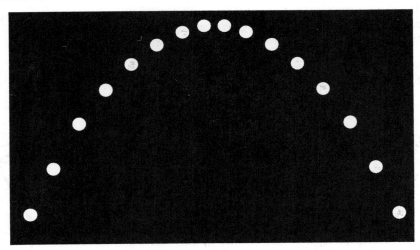

The orbits of planets such as the Earth and Mars are ellipses. Satellites also usually follow elliptical orbits.

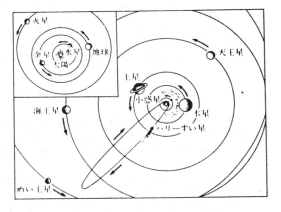

Inset labels, inner to outer:
    Mercury
    Venus
    Earth
    Mars

Larger figure labels, inner to outer:
    Halley's Comet [the flattened ellipse]
    Mars
    Jupiter
    Saturn
    Uranus
    Neptune
    Pluto

When ripples are formed on a water surface, quadratic curves sometimes appear as the set of points where the ripples interfere with each other. The figure to the right is one example.

Foci and Quadratic Curves

At the surface of a mirror made by revolving a quadratic curve about the vertex, when light shines to the mirror from focus F or F', the light follows the paths shown in the figure below.

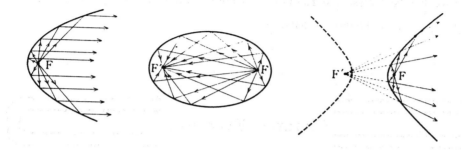

Exercises

1. Find the equation of an ellipse in which the foci are two points $F(2, 0)$ and $F'(-2, 0)$ and the length of the major axis is 10.

2. Find the vertices, foci, and asymptotes of the following hyperbolas. Then sketch the curves.

   (1) $\dfrac{x^2}{36} - \dfrac{y^2}{4} = 1$ (2) $x^2 - y^2 + 2y = 0$

3. Find the equation of a hyperbola in which the foci are $F(\sqrt{5}, 0)$ and $F'(-\sqrt{5}, 0)$ and the asymptotes are two lines $y = \pm 2x$.

4. Take P as the point which divides line segment AB of length 5 internally at a ratio of $2 : 3$. If the endpoints A and B of line segment AB move along the x- and y-axes:

   (1) Take $A(u, 0)$, $B(0, v)$, and $P(x, y)$, and show that $u = \dfrac{5}{3}x$, $v = \dfrac{5}{2}y$.

   (2) Find the locus of points P.

5. Find the locus of points P such that the ratio of its distances from point $F(4, 0)$ and the straight line $x = 1$ is $2 : 1$.

6. How will the number of common points shared by the ellipse $3x^2 + y^2 = 3$ and the straight line $y = mx + 3$ change according to the value of m?

7. Prove that the necessary and sufficient condition for the line $y = mx$ and the hyperbola $\frac{x^2}{a^2} - \frac{y^2}{b^2} = 1$ to intersect is $|m| < \frac{b}{a}$.

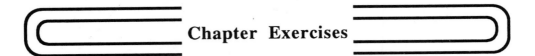

Chapter Exercises

A

1. Sketch the following curves:

   (1) $y^2 - 4y = 4x$ (2) $y^2 - 4y = 4x^2$ (3) $y^2 - 4y = -4x^2$

2. Find the equation of an ellipse in which the foci are $(\sqrt{3}, 0)$ and $(-\sqrt{3}, 0)$ and which passes through the point $(2, -1)$.

3. Given a circle with a radius of a and a center at the fixed point F. Take one fixed point F' inside this circle, and take Q as a point moving along the circumference. Take P as the point where the perpendicular bisector of QF' intersects QF. Then the locus of P is an ellipse with F and F' as its foci. Prove this statement.

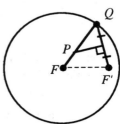

4. In Problem 3, prove that the locus of point P is a hyperbola with F and F' as its foci if fixed point F' lies outside of circle F.

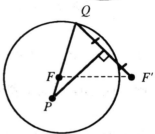

5. Take B and B' as the points at which the ellipse $\dfrac{x^2}{9} + \dfrac{y^2}{4} = 1$ intersects the y-axis. Take Q and Q' as the points at which the x-axis intersects lines BP and $B'P$, which connect B and B' with point P, distinct from B and B', on the circumference.

   (1) Express the coordinates of Q and Q' in terms of x_1 and y_1 by assuming $P(x_1, y_1)$.

   (2) Prove that $OQ \cdot OQ'$ is a constant.

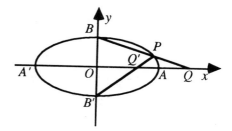

B

1. Find the equations of the following parabolas:

   (1) focus $(2, 3)$, directrix $x = -2$

   (2) focus $(-1, 2)$, directrix $x = 3$

2. Find the range of m such that the straight line $y = mx + 1$ and the hyperbola $x^2 - y^2 = 4$ have a common point.

3. If the product of the slopes of two straight lines connecting one point P and two fixed points $(a, 0)$ and $(-a, 0)$ is a non-zero constant k, P lies on an ellipse or a hyperbola. Prove this statement.

4. The upper figure at the right illustrates the principle of the elliptical compass shown in the lower figure. Let $AP = a$ and $BP = b$, and assume that A moves along the y-axis and B moves along the x-axis. Prove that the locus of P is an ellipse, using the fact that

$$PB^2 = PQ^2 + BQ^2$$

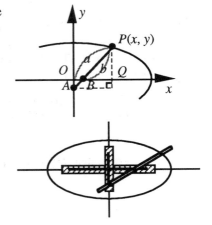

5. Take A and A' as the points at which the ellipse $\dfrac{x^2}{16} + \dfrac{y^2}{9} = 1$ intersects the x-axis, and take P and P' as the points at which an arbitrary line $x = x_1$ parallel to the y-axis intersects this ellipse, provided that $x_1 \neq 0$.

   (1) Find the equations of lines $A'P$ and $P'A$ by assuming $P(x_1, y_1)$ and $P'(x_1, -y_1)$.

   (2) Prove that the equation of the locus of points R at which the two lines $A'P$ and $P'A$ intersect is $\dfrac{x^2}{16} - \dfrac{y^2}{9} = 1$.

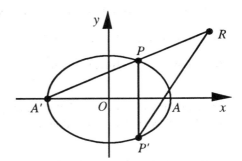

CHAPTER 2

VECTORS IN THE PLANE

SECTION 1. VECTORS AND OPERATIONS INVOLVING THEM
SECTION 2. APPLICATIONS OF VECTORS

The concept of the vector was born in physics. Originally it was considered a quantity with magnitude and direction, such as force, velocity, and acceleration in dynamics.

The term "vectors" originally denoted "vectors under a constraint." In dealing with a force vector, for example, the point at which a force is applied is significant. When we eliminate the constraint, the concept of a "free vector" arises. Although vectors in present-day geometry are usually understood to be free vectors, mathematics was to undergo a long period of development before this concept was clarified mathematically and vector operations were established. In this chapter you will learn how useful the concept of a vector is in the most basic area of plane geometry.

Note that the concept of the vector has now been generalized, and it is the fundamental concept of an important field of mathematics, linear algebra.

VECTORS AND OPERATIONS INVOLVING THEM

The Meaning of a Vector

The motion of a point moving from A to B in a plane can be represented by a line segment AB with an arrow, as in the figure below. When the direction of the line segment is specified by an arrow, we call the segment a **directed line segment**, and A is the **initial point** and B is the **end point**.

Motion from B to A can be represented by a line segment with the opposite arrow, as in the figure to the right.

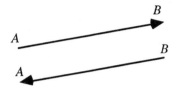

Example We often express wind conditions in an area with a phrase such as

"a SW wind at 5 m/sec"

This means the wind is blowing from the southwest to the northeast at a velocity of 5 m/sec.

The direction and velocity of the wind can be represented by a directed line segment AB of length 5 from point A in the direction of the northeast. In this case, the direction and length of the directed line segment are important, but the base point can be anywhere.

Therefore, with respect to direction and velocity, directed line segments AB, CD, EF, ..., which all have the same direction and the same length in the figure to the right, can be regarded as equivalent.

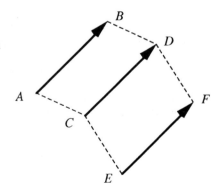

2 VECTORS IN THE PLANE

When we consider only the direction and length of a directed line segment AB in a plane and neglect its location, as in the above Example, we refer to it as a **vector** in the plane and designate it by $\overrightarrow{AB}$.

We refer to the length of line segment AB as the **magnitude** or length of vector $\overrightarrow{AB}$ and designate it by $|\overrightarrow{AB}|$. Therefore,

$$|\overrightarrow{AB}| = AB.$$

Vectors are defined by their magnitude and direction, so the fact that $\overrightarrow{AB}$ and $\overrightarrow{CD}$ are **equal** means that their magnitude and direction are the same. In other words, directed line segment AB can be placed on directed line segment CD by translation.

When $\overrightarrow{AB}$ and $\overrightarrow{CD}$ are equal, we write

$$\overrightarrow{AB} = \overrightarrow{CD}.$$

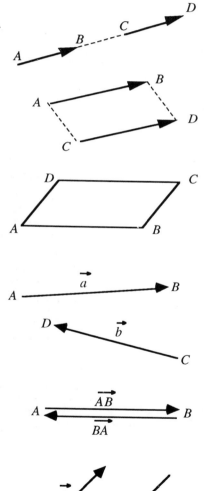

Problem 1 State which vectors are equal to $\overrightarrow{BC}$ in parallelogram $ABCD$ to the right. State which vectors are equal to $\overrightarrow{CD}$.

Vectors are also designated by single letters with an arrow such as $\vec{a}$ and $\vec{b}$,

$$\vec{a} = \overrightarrow{AB}, \ \vec{b} = \overrightarrow{CD}.$$

$\overrightarrow{AB}$ and $\overrightarrow{BA}$ have the same magnitude but opposite direction. A vector that has the same magnitude as $\vec{a}$ but the opposite direction is called the **inverse vector** and is designated by $-\vec{a}$. Therefore,

$$-\overrightarrow{AB} = \overrightarrow{BA}.$$

36 2 VECTORS IN THE PLANE

Problem 2 Which vectors are equal in the figure to the right? Which are inverse vectors to each other?

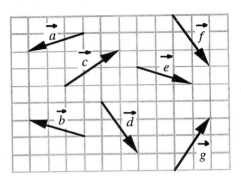

 Addition and Subtraction of Vectors

Addition of Vectors

Example A man walked 5 km to the east from point A and reached point B, and then walked 5 km to the north of point B and reached point C. This man is located at a point $5\sqrt{2}$ km to the northeast of point A. Thus, the result of the movements of $\overrightarrow{AB}$ and $\overrightarrow{BC}$, taken together, was the movement $\overrightarrow{AC}$.

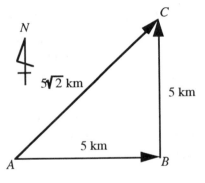

In the above Example, the result of movements $\overrightarrow{AB}$ and $\overrightarrow{BC}$, taken together, is the movement $\overrightarrow{AC}$. In this way, we will define the addition of vectors by considering successive movements together, that is, the composition of movements is the sum.

Take $\vec{AB}$ as a vector equal to $\vec{a}$, and $\vec{BC}$ as a vector equal to $\vec{b}$. Now let us define vector $\vec{c} = \vec{AC}$ as the **sum** of $\vec{a}$ and $\vec{b}$ and write it as

$$\vec{a} + \vec{b} = \vec{c}.$$

This definition does not depend in any way upon the location of the initial point A.

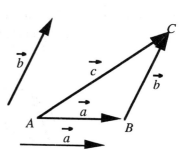

Problem 1   Sketch a diagram to illustrate the following movements by vectors.

Move 4 km to the west of A, 3 km to the north, 5 km to the east, and 6 km to the south to reach B.

Problem 2   Show the sums $\vec{a} + \vec{b}$, $\vec{c} + \vec{d}$, and $\vec{e} + \vec{f}$ in the figure below.

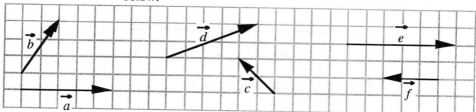

Problem 3   Check that $|\vec{a} + \vec{b}| \leq |\vec{a}| + |\vec{b}|$.

We can formulate the following fundamental properties for the addition of vectors.

---

(1) $\vec{a} + \vec{b} = \vec{b} + \vec{a}$ Commutative law

(2) $(\vec{a} + \vec{b}) + \vec{c} = \vec{a} + (\vec{b} + \vec{c})$ Associative law

---

$(\vec{a} + \vec{b}) + \vec{c}$ or $\vec{a} + (\vec{b} + \vec{c})$ can be written as $\vec{a} + \vec{b} + \vec{c}$, omitting the parentheses.

38 2 VECTORS IN THE PLANE

Problem 4 Check that the commutative law holds for the addition of vectors by using the figure on the left below. Then check that the associative law holds by using the figure on the right.

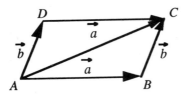
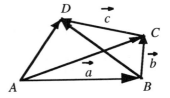

If we take $\vec{a} = \overrightarrow{AB}$, then $-\vec{a} = \overrightarrow{BA}$. Therefore,

$$\vec{a} + (-\vec{a}) = \overrightarrow{AB} + \overrightarrow{BA} = \overrightarrow{AA}.$$

$\overrightarrow{AA}$ is not a directed line segment, but let us consider it as a vector with a magnitude of 0 and designate it as $\vec{0}$. $\vec{0}$ is called a **zero vector**.

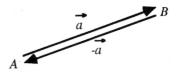

Properties of a Zero Vector

(1) The magnitude of a zero vector is $|\vec{0}| = 0$.

(2) $\vec{a} + \vec{0} = \vec{a}$

(3) $\vec{a} + (-\vec{a}) = \vec{0}$

Problem 5 Prove the equality $\overrightarrow{AB} + \overrightarrow{BC} + \overrightarrow{CA} = \vec{0}$.

Subtraction of Vectors

Given two vectors $\vec{a}$ and $\vec{b}$, $\vec{a} + (-\vec{b})$ can be expressed as

$$\vec{a} - \vec{b}$$

and is referred to as the **difference** of $\vec{a}$ and $\vec{b}$.

Problem 6 If we take $\vec{a} - \vec{b}$ as $\vec{x}$, then $\vec{x}$ satisfies

$$\vec{b} + \vec{x} = \vec{a}.$$

Check this.

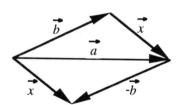

Problem 7 Find $\vec{AB} - \vec{AC}$ and $\vec{AC} - \vec{AB}$ in the figure to the right.

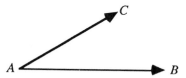

Problem 8 Check that $-(\vec{a} + \vec{b}) = -\vec{a} - \vec{b}$.

   Multiplying a Vector by a Real Number

Multiplying a Vector by a Real Number

It is customary to consider that doubling a movement of 2 km to the east results in a movement of 4 km to the east, and that multiplying -1.5 times a movement of 2 km to the east yields a movement of 3 km to the west.

Accordingly, given a vector $\vec{a}$ and a real number m, we can define $m\vec{a}$, m times $\vec{a}$, in the following way:

> (i) If $\vec{a}$ is not equal to $\vec{0}$:
>
> (1) If $m > 0$, $m\vec{a}$ is a vector with the same direction as $\vec{a}$ and a length of $m|\vec{a}|$.
>
> (2) If $m < 0$, $m\vec{a}$ is a vector with the opposite direction from $\vec{a}$ and a length of $|m||\vec{a}|$.
>
> (3) If $m = 0$, $m\vec{a}$ is a zero vector $\vec{0}$.
>
> (ii) If $\vec{a}$ is equal to $\vec{0}$, $m\vec{a}$ is a zero vector $\vec{0}$.

From the definition above, we have the following special cases:

$$1\vec{a} = \vec{a}, \qquad (-1)\vec{a} = -\vec{a}$$

$$0\vec{a} = \vec{0}, \qquad m\vec{0} = \vec{0}$$

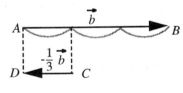

Note: For $m = \dfrac{1}{k}$, $\dfrac{1}{k}\vec{a}$ is sometimes written as $\dfrac{\vec{a}}{k}$.

Parallel Vectors

When two vectors $\vec{a}$ and $\vec{b}$, not equal to $\vec{0}$, have the same or opposite direction, $\vec{a}$ and $\vec{b}$ are said to be **parallel**, and are written as $\vec{a} \parallel \vec{b}$.

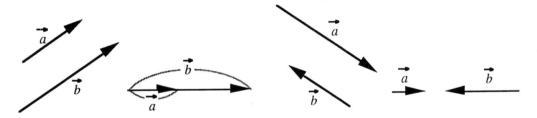

From the definition of how vectors are multiplied by real numbers and the definition of parallel vectors, we can see that two vectors not equal to $\vec{0}$ are parallel if and only if one is created by multiplying the other by a real number.

For $\vec{a} \neq \vec{0}$ and $\vec{b} \neq \vec{0}$,

$$\vec{a} \parallel \vec{b} \leftrightarrow \vec{a} = m\vec{b}.$$

Problem 1 If $\vec{AB} = m\vec{AC}$, then the three points A, B, and C lie on a single line. Prove this statement.

The following fundamental properties hold when vectors are multiplied by a real number.

---

(1) $(mn)\vec{a} = m(n\vec{a})$ Associative law

(2) $(m+n)\vec{a} = m\vec{a} + n\vec{a}$ Distributive law I

(3) $m(\vec{a} + \vec{b}) = m\vec{a} + m\vec{b}$ Distributive law II

$(mn)\vec{a}$ or $m(n\vec{a})$ can be written as $mn\,\vec{a}$, omitting the parentheses. The fact that the associative law and distributive law I hold is immediately clear from the two figures to the right.

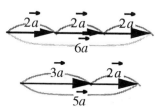

Problem 2
Check that distributive law II holds, using the figure to the right.

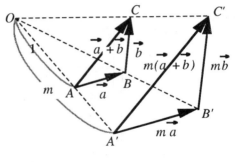

Problem 3
Perform the following calculations:

(1) $3\vec{a} + 4\vec{a} - \vec{a}$

(2) $3(\vec{u} + 2\vec{v}) - 3(\vec{u} - 4\vec{v})$

Problem 4
Take L, M, and N as the midpoints of sides BC, CA, and AB of $\triangle ABC$. Express the following vectors in terms of $\vec{a}$ and $\vec{b}$ by setting $\vec{AB} = \vec{a}$ and $\vec{AC} = \vec{b}$.

$\vec{NL}$, $\vec{LM}$, $\vec{MN}$, $\vec{AL}$, $\vec{BM}$, $\vec{CN}$

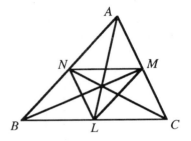

A vector with a length of 1 is called a **unit vector**.

If we take an arbitrary vector $\vec{a}$, not equal to $\vec{0}$, then

$$\vec{e} = \frac{\vec{a}}{|\vec{a}|}$$

is a unit vector with the same direction as $\vec{e}$.

Problem 5
Check the above formula.

2 VECTORS IN THE PLANE

  Components of Vectors

Vectors and the Coordinate Axes

Take orthogonal coordinate axes on a plane, with O as the origin, as the x-axis and y-axis.

In this case, a unit vector that has the same direction as the positive ray of the x-axis is called the **basic vector of the x-axis**, and is designated by $\vec{e}_1$. We can define the **basic vector of the y-axis** analogously and designate it by $\vec{e}_2$.

Take points $E_1\,(1, 0)$ and $E_2\,(0, 1)$ on the x-axis and the y-axis, and then

$$\overrightarrow{OE}_1 = \vec{e}_1, \quad \overrightarrow{OE}_2 = \vec{e}_2$$

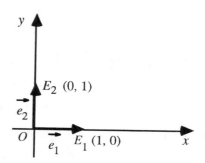

$$|\vec{e}_1| = 1, \quad |\vec{e}_2| = 1.$$

For a given vector $\vec{a}$, take point P such that

$$\vec{a} = \overrightarrow{OP}.$$

Take PP_1 and PP_2 as the perpendicular lines from P to the x-axis and the y-axis. Then we can see from the figure that

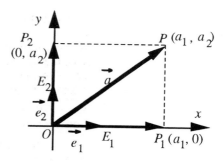

$$\vec{a} = \overrightarrow{OP} = \overrightarrow{OP}_1 + \overrightarrow{P_1P} = \overrightarrow{OP}_1 + \overrightarrow{OP}_2. \qquad (1)$$

In (1), if we take the coordinates of P as (a_1, a_2), then the coordinates of P_1 and P_2 are $(a_1, 0)$ and $(0, a_2)$. From the definition of how vectors are multiplied by a real number, $\overrightarrow{OP}_1$ and $\overrightarrow{OP}_2$ can be expressed as

$$\overrightarrow{OP}_1 = a_1 \overrightarrow{OE}_1 = a_1 \vec{e}_1, \quad \overrightarrow{OP}_2 = a_2 \overrightarrow{OE}_2 = a_2 \vec{e}_2.$$

44 2 VECTORS IN THE PLANE

Substituting these expressions into (1), $\vec{a}$ can be expressed using the basic vectors $\vec{e}_1$ and $\vec{e}_2$ as

$$\vec{a} = a_1 \vec{e}_1 + a_2 \vec{e}_2.$$

These values a_1 and a_2 are called the **x-component** and the **y-component** of $\vec{a}$.

The magnitude of vector $\vec{a}$ is

$$|\vec{a}| = |\overrightarrow{OP}| = \sqrt{a_1^2 + a_2^2}$$

Demonstration 1   Find the x- and y-components of vector $\overrightarrow{AB}$ from point A (-2, 1) to point B (3, 4), and express $\overrightarrow{AB}$ in terms of basic vectors. Then find $|\overrightarrow{AB}|$.

[Solution] Take point P satisfying $\overrightarrow{AB} = \overrightarrow{OP}$, and then the coordinates of P are

$$(3 + 2, 4 - 1) = (5, 3)$$

as you can see clearly from the figure. Therefore, the x-component of $\overrightarrow{AB} = \overrightarrow{OP}$ is 5 and the y-component is 3. Thus, $\overrightarrow{AB}$ can be expressed in terms of basic vectors as

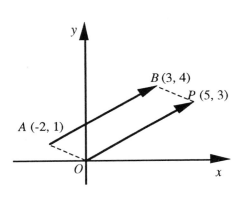

$$\overrightarrow{AB} = 5\vec{e}_1 + 3\vec{e}_2.$$

Furthermore,

$$|\overrightarrow{AB}| = \sqrt{5^2 + 3^2} = \sqrt{34}.$$

Problem 1   Given points A (-2, -6), B (3, 1), and C (3, 4) express $\overrightarrow{AB}$, $\overrightarrow{BC}$, $\overrightarrow{CA}$ in terms of basic vectors. Then calculate the magnitude of each vector.

2 VECTORS IN THE PLANE   45

Problem 2 Express the vectors $\vec{a}$, $\vec{b}$, $\vec{c}$, and $\vec{d}$ in the figure to the right in terms of basic vectors.

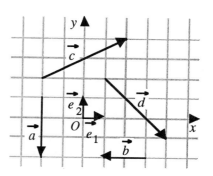

In general, given $A(x, y)$ and $B(x', y')$, the x-component a_1 and the y-component a_2 of vector $\overrightarrow{AB}$ are given by the following expressions:

$$a_1 = x' - x, \quad a_2 = y' - y.$$

If the x- and y-components of vector $\vec{a}$ in a plane are a_1 and a_2, or $\vec{a} = a_1 \vec{e}_1 + a_2 \vec{e}_2$, vector $\vec{a}$ can be written simply as vector (a_1, a_2). That is,

$$\vec{a} = (a_1, a_2).$$

This method of writing a vector $\vec{a}$ is called the **component representation** with respect to the coordinate axes.

$\vec{a} = a_1 \vec{e}_1 + a_2 \vec{e}_2$	Representation in terms of basic vectors
$\vec{a} = (a_1, a_2)$	Component representation

As special cases, the component representations of the basic vectors and the zero vector are

$$\vec{e}_1 = (1, 0), \quad \vec{e}_2 = (0, 1), \quad \vec{0} = (0, 0).$$

The relation between the components of identical vectors is given below:

$$(a_1, a_2) = (b_1, b_2) \leftrightarrow a_1 = b_1 \text{ and } a_2 = b_2.$$

Problem 3 Represent $\overrightarrow{AB}$ and $\overrightarrow{BA}$ in Demonstration 1 by means of components.

2 VECTORS IN THE PLANE

Problem 4 Sketch the following vectors, taking the origin as the initial point:

(1) $\vec{a} = 2e_1 - 3e_2$ (2) $\vec{b} = 5e_2$

(3) $\vec{c} = \left(-\dfrac{5}{2}, -3\right)$ (4) $\vec{d} = (4, 0)$

Calculation by Means of Components

Let's consider how component representation can help us perform calculations involving vectors.

Demonstration 2 Represent $\vec{a} + \vec{b}$ by means of components, if $\vec{a} = (a_1, a_2)$ and $\vec{b} = (b_1, b_2)$.

[Solution] Since $\vec{a} = (a_1, a_2)$, $\vec{a} = a_1 \vec{e}_1 + a_2 \vec{e}_2$.

Since $\vec{b} = (b_1, b_2)$, $\vec{b} = b_1 \vec{e}_1 + b_2 \vec{e}_2$

$$\vec{a} + \vec{b} = (a_1 \vec{e}_1 + a_2 \vec{e}_2) \; (b_1 \vec{e}_1 + b_2 \vec{e}_2)$$

$$= (a_1 + b_1) \vec{e}_1 + (a_2 + b_2) \vec{e}_2.$$

Therefore, the component representation of $\vec{a} + \vec{b}$ is

$$\vec{a} + \vec{b} = (a_1 + b_1, a_2 + b_2).$$

From Demonstration 2 we know that the equality

$$(a_1, a_2) + (b_1, b_2) = (a_1 + b_1, a_2 + b_2)$$

holds. This is a formula for finding the sum of vectors by means of component representation. We can derive similar formulas for the difference of two vectors, and for the product of a vector and a real number.

Calculation by Means of Components

(1) $(a_1, a_2) + (b_1, b_2) = (a_1 + b_1, a_2 + b_2)$

(2) $(a_1, a_2) - (b_1, b_2) = (a_1 - b_1, a_2 - b_2)$

(3) $m(a_1, a_2) = (ma_1, ma_2)$

Problem 5 Derive formulas (2) and (3) above as in Demonstration 2.

Problem 6 Find the component representations of the following vectors $\vec{u}$, $\vec{v}$, and $\vec{w}$, if $\vec{a} = (2, -3)$, $\vec{b} = (-1, 2)$, and $\vec{c} = (5, 0)$.

(1) $\vec{u} = \vec{a} + \vec{b} + \vec{c}$ (2) $\vec{v} = -3\vec{a} + 2\vec{b} + \vec{c}$

(3) $\vec{w} = 3(\vec{a} + \vec{b}) - 2(\vec{b} - \vec{c})$

The magnitude of a vector given by component representation can be found from the following formula.

Magnitude of a Vector

If $\vec{a} = (a_1, a_2)$, then $|\vec{a}| = \sqrt{a_1^2 + a_2^2}$.

Problem 7 Find the magnitude of vectors $\vec{a}$, $\vec{b}$ and $\vec{c}$ in Problem 6.

Inner Product of Vectors

Take two vectors $\vec{a}$ and $\vec{b}$ not equal to $\vec{0}$. Take point O as their common initial point, then take two points A and B satisfying

$$\overrightarrow{OA} = \vec{a}, \ \overrightarrow{OB} = \vec{b}.$$

In this case $\angle AOB = \theta$ is said to be the angle formed by $\vec{a}$ and $\vec{b}$ (provided that $0° \leq \theta \leq 180°$). Moreover,

$$|\vec{a}||\vec{b}|\cos\theta$$

is called the **inner product** of $\vec{a}$ and $\vec{b}$, and is designated by the symbol $\vec{a} \cdot \vec{b}$.

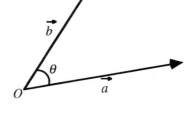

Inner Product of Vectors

Take θ as the angle formed by two vectors $\vec{a}$ and $\vec{b}$, and then

$$\vec{a} \cdot \vec{b} = |\vec{a}||\vec{b}|\cos\theta.$$

For $\vec{a} = \vec{0}$ or $\vec{b} = \vec{0}$, we will stipulate that

$$\vec{a} \cdot \vec{b} = 0.$$

Example 1 In an equilateral triangle ABC with sides of length 2, take

$$\vec{CA} = \vec{a}, \ \vec{CB} = \vec{b}.$$

Then the angle formed by $\vec{a}$ and $\vec{b}$ is $60°$, and therefore,

$$\vec{a} \cdot \vec{b} = |\vec{a}||\vec{b}|\cos 60°$$
$$= 2 \times 2 \times \frac{1}{2} = 2.$$

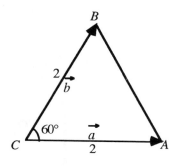

From the definition of an inner product, it is clear that

$$\vec{a} \cdot \vec{b} = \vec{b} \cdot \vec{a}.$$

Since

$$-1 \leq \cos \theta \leq 1$$

for any vectors $\vec{a}$ and $\vec{b}$,

$$|\vec{a} \cdot \vec{b}| \leq |\vec{a}||\vec{b}|.$$

Moreover, if $\vec{a} = \vec{b}$, then $\theta = 0°$. Thus,

$$\cos \theta = 1.$$

Therefore,

$$\vec{a} \cdot \vec{a} = |\vec{a}|^2.$$

Problem 1 Show that the following equalities hold by virtue of the definition of an inner product.

For $\theta = 0°$, $\vec{a} \cdot \vec{b} = |\vec{a}||\vec{b}|$.

For $\theta = 180°$, $\vec{a} \cdot \vec{b} = -|\vec{a}||\vec{b}|$.

2 VECTORS IN THE PLANE

If two vectors $\vec{a}$ and $\vec{b}$ form a right angle, $\vec{a}$ and $\vec{b}$ are said to be **orthogonal** or **perpendicular** to each other; this relation is designated by $\vec{a} \perp \vec{b}$. In this case, since

$$\cos \theta = 0,$$

the following implication holds.

The Inner Product of Orthogonal Vectors

$$\vec{a} \perp \vec{b} \leftrightarrow \vec{a} \cdot \vec{b} = 0$$

We can assume that $\vec{0}$ is perpendicular to any other vector.

Example 2 Since the basic vectors e_1 and e_2 are unit vectors which are orthogonal to each other,

$$\vec{e}_1 \cdot \vec{e}_1 = \vec{e}_2 \cdot \vec{e}_2 = 1$$

$$\vec{e}_1 \cdot \vec{e}_2 = 0.$$

Problem 2 Find the following inner products in the right triangle illustrated at the right.

(1) $\vec{AB} \cdot \vec{AC}$ (2) $\vec{CA} \cdot \vec{CB}$

(3) $\vec{AB} \cdot \vec{BC}$ (4) $\vec{AB} \cdot \vec{CA}$

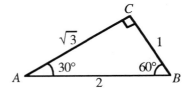

Problem 3 Isosceles right triangle ABC has $\angle A$ as the right angle. Take a as the length of AB, draw perpendicular AM from A to side BC. Find the following inner products:

(1) $\vec{AB} \cdot \vec{CB}$ (2) $\vec{BA} \cdot \vec{CA}$ (3) $\vec{AM} \cdot \vec{BA}$

Inner Products and Component Representations

Assume that the component representations of two vectors $\vec{a}$ and $\vec{b}$ are

$$\vec{a} = (a_1, a_2), \quad \vec{b} = (b_1, b_2).$$

Then assume that

$$\overrightarrow{OA} = \vec{a}, \quad \overrightarrow{OB} = \vec{b}$$

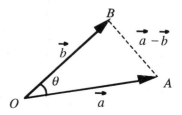

as in the figure to the right, and take $\angle AOB = \theta$. From the cosine theorem,

$$2OA \cdot OB \cos\theta = OA^2 + OB^2 - BA^2. \tag{1}$$

Since

$$OA^2 = |\vec{a}|^2 = a_1^2 + a_2^2$$

$$OB^2 = |\vec{b}|^2 = b_1^2 + b_2^2$$

$$BA^2 = |\vec{a} - \vec{b}|^2 = (a_1 - b_1)^2 + (a_2 - b_2)^2,$$

we can determine that in (1),

$$\text{Right side} = OA^2 + OB^2 - BA^2 = 2(a_1 b_1 + a_2 b_2)$$

$$\text{Left side} = 2OA \cdot OB \cos\theta = 2|\vec{a}||\vec{b}|\cos\theta = 2(\vec{a} \cdot \vec{b}).$$

Therefore,

$$\vec{a} \cdot \vec{b} = a_1 b_1 + a_2 b_2.$$

Thus, we can formulate the following generalization.

Components and the Inner Product

If $\vec{a} = (a_1, a_2)$ and $\vec{b} = (b_1, b_2)$,

$$\vec{a} \cdot \vec{b} = a_1 b_1 + a_2 b_2.$$

As a special case,

$$\vec{a} \perp \vec{b} \leftrightarrow a_1 b_1 + a_2 b_2 = 0.$$

In general, if we take θ as the angle formed by $\vec{a} = (a_1, a_2)$ and $\vec{b} = (b_1, b_2)$, then

$$\cos\theta = \frac{\vec{a} \cdot \vec{b}}{|\vec{a}||\vec{b}|} = \frac{a_1 b_1 + a_2 b_2}{\sqrt{a_1^2 + a_2^2}\sqrt{b_1^2 + b_2^2}}.$$

Problem 4 Find the inner products of vectors $\vec{a}$ and $\vec{b}$.

(1) $\vec{a} = (-2, 3)$, $\vec{b} = (5, 4)$

(2) $\vec{a} = 3\vec{e}_1 + 5\vec{e}_2$, $\vec{b} = -4\vec{e}_1 + \vec{e}_2$

Problem 5 Show that $\vec{a} \perp \vec{b}$ for vectors $\vec{a} = (3, -4)$ and $\vec{b} = (8, 6)$.

Problem 6 Find the angle θ formed by vectors $\vec{a}$ and $\vec{b}$.

(1) $\vec{a} = (-3, 0)$, $\vec{b} = (-1, \sqrt{3})$

(2) $\vec{a} = (2, 1)$, $\vec{b} = (3, -6)$

Demonstration 1 For any vectors $\vec{a}$, $\vec{b}$ and $\vec{c}$, prove that the following formula holds:

$$\vec{a} \cdot (\vec{b} + \vec{c}) = \vec{a} \cdot \vec{b} + \vec{a} \cdot \vec{c}.$$

[**Proof**]: Assume that the component representations of $\vec{a}$, $\vec{b}$ and $\vec{c}$ are:

$$\vec{a} = (a_1, a_2), \quad \vec{b} = (b_1, b_2), \quad \vec{c} = (c_1, c_2).$$

Then

$$\vec{b} + \vec{c} = (b_1 + c_1, b_2 + c_2).$$

Therefore,

$$\vec{a} \cdot (\vec{b} + \vec{c}) = a_1(b_1 + c_1) + a_2(b_2 + c_2)$$
$$= (a_1 b_1 + a_2 b_2) + (a_1 c_1 + a_2 c_2)$$
$$= \vec{a} \cdot \vec{b} + \vec{a} \cdot \vec{c}.$$

Analogously, the following formulas also hold for an arbitrary real number m.

$$(\vec{a} + \vec{b}) \cdot \vec{c} = \vec{a} \cdot \vec{c} + \vec{b} \cdot \vec{c}$$

$$\vec{a} \cdot (m\vec{b}) = m(\vec{a} \cdot \vec{b})$$

$$(m\vec{a}) \cdot \vec{b} = m(\vec{a} \cdot \vec{b})$$

Problem 7 Prove the above formulas.

Problem 8 Prove the following equalities:

$$\vec{a} \cdot (\vec{b} - \vec{c}) = \vec{a} \cdot \vec{b} - \vec{a} \cdot \vec{c}$$

$$(\vec{a} - \vec{b}) \cdot \vec{c} = \vec{a} \cdot \vec{c} - \vec{b} \cdot \vec{c}$$

2 VECTORS IN THE PLANE

Demonstration 2 Find the value of

$$|\vec{a} + 2\vec{b}|$$

if $|\vec{a}| = 2$, $|\vec{b}| = 3$, $\vec{a} \cdot \vec{b} = 4$.

[Solution]

$$|\vec{a} + 2\vec{b}|^2 = (\vec{a} + 2\vec{b}) \cdot (\vec{a} + 2\vec{b})$$

$$= \vec{a} \cdot (\vec{a} + 2\vec{b}) + 2\vec{b} \cdot (\vec{a} + 2\vec{b})$$

$$= \vec{a} \cdot \vec{a} + 2\vec{a} \cdot \vec{b} + 2\vec{b} \cdot \vec{a} + 4\vec{b} \cdot \vec{b}$$

$$= |\vec{a}|^2 + 4\vec{a} \cdot \vec{b} + 4|\vec{b}|^2$$

$$= 2^2 + 4 \times 4 + 4 \times 3^2 = 56$$

Therefore,

$$|\vec{a} + 2\vec{b}| = \sqrt{56} = 2\sqrt{14}.$$

Problem 9 Find the value of $|2\vec{a} - 3\vec{b}|$ for $\vec{a}$ and $\vec{b}$ in Demonstration 2.

Exercises

1. Given vectors $\vec{u}$ and $\vec{v}$ as in the figure to the right, calculate the following expressions, and show the results on the figure.

   (1) $6\vec{u} - 5\vec{v} - 4\vec{u} + 2\vec{v}$

   (2) $7(\vec{u} - 2\vec{v}) - 4(2\vec{u} - 3\vec{v})$

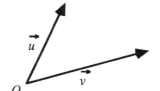

2. Find real numbers k and l satisfying $\vec{c} = k\vec{a} + l\vec{b}$, if $\vec{a} = (-2, 3)$, $\vec{b} = (1, -4)$ and $\vec{c} = (8, -17)$.

3. Prove the following equalities:

   (1) $(4\vec{a} + 3\vec{b}) \cdot (4\vec{a} - 3\vec{b}) = 16|\vec{a}|^2 - 9|\vec{b}|^2$

   (2) $|\vec{a} + \vec{b}|^2 - |\vec{a} - \vec{b}|^2 = 4\vec{a} \cdot \vec{b}$

4. Find the angle θ formed by vectors $\vec{a}$ and $\vec{b}$ in each case:

   (1) $|\vec{a}| = 3,\ |\vec{b}| = 4,\ \vec{a} \cdot \vec{b} = 6$

   (2) $|\vec{a}| = |\vec{b}| = \vec{a} \cdot \vec{b} = \sqrt{2}$

5. Given vectors $\vec{a} = (2, 1)$ and $\vec{b} = (-1, 2)$. Specify the value of real number x such that the vectors $4x\vec{a} + \vec{b}$ and $x\vec{a} - 3\vec{b}$ will be perpendicular.

6. Prove that for two vectors $\vec{a}$ and $\vec{b}$ not equal to 0, if $|\vec{a} + \vec{b}| = |\vec{a} - \vec{b}|$, then $\vec{a} \perp \vec{b}$.

7. Take $\vec{e}$ as the unit vector parallel to line l through point O, and take B as the point at which a perpendicular line from point A intersects l. Show that $|\vec{OA} \cdot \vec{e}| = OB$.

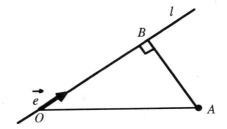

2 APPLICATIONS OF VECTORS

1 Position Vectors

If a fixed point O is defined on a plane, for any arbitrary point P we can define a vector

$$\overrightarrow{OP} = \vec{p}. \qquad (1)$$

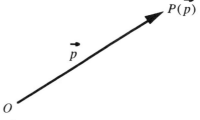

This vector $\vec{p}$ is called the **position vector** of point P with point O as the origin.

The fact that the position vector of point P is $\vec{p}$ is designated by $P(\vec{p})$.

Conversely, given vector $\vec{p}$, we can identify point P with $\vec{p}$ as its position vector from (1).

Let's express vector $\overrightarrow{AB}$ in a plane by means of position vectors of A and B.

Take $\vec{a}$ and $\vec{b}$ as the position vectors of points A and B, respectively. Then

$$\overrightarrow{OA} = \vec{a}, \ \overrightarrow{OB} = \vec{b}.$$

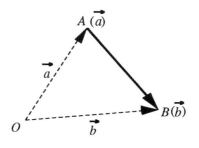

Substituting these expressions into

$$\overrightarrow{AB} = \overrightarrow{OB} - \overrightarrow{OA}$$

we obtain

$$\overrightarrow{AB} = \vec{b} - \vec{a}.$$

In other words, vector $\overrightarrow{AB}$ is equal to the difference of the position vector of the initial point A and the position vector of the end point B.

2 VECTORS IN THE PLANE

Problem 1   Express the condition such that quadrilateral $ABCD$ is a parallelogram in terms of the position vectors $\vec{a}$, $\vec{b}$, $\vec{c}$, and $\vec{d}$ of each vertex A, B, C, and D of the quadrilateral.

Demonstration 1   Prove that the position vector $\vec{c}$ of point C, which internally divides line segment AB connecting two points $A(\vec{a})$ and $B(\vec{b})$ at a ratio of $m:n$, can be expressed in the following form:

$$\vec{c} = \frac{m\vec{b} + n\vec{a}}{m+n}.$$

[Proof]   If we represent $\overrightarrow{AC}$ and $\overrightarrow{AB}$ by position vectors, then

$$\overrightarrow{AC} = \vec{c} - \vec{a}$$

$$\overrightarrow{AB} = \vec{b} - \vec{a}.$$

From the figure, it is clear that

$$\overrightarrow{AC} = \frac{m}{m+n} \overrightarrow{AB}.$$

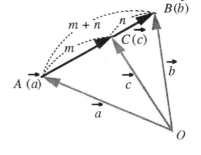

Therefore,

$$\vec{c} - \vec{a} = \frac{m}{m+n}(\vec{b} - \vec{a}).$$

Transforming this equality, we obtain

$$\vec{c} = \frac{m\vec{b} + n\vec{a}}{m+n}.$$

As a special case, **the position vector of the midpoint of line segment AB is** $\dfrac{\vec{a} + \vec{b}}{2}$.

Problem 2   Prove that the position vector $\vec{d}$ of point D, which externally divides line segment AB connecting the two points $A(\vec{a})$ and $B(\vec{b})$, can be expressed in the following form:

$$\vec{d} = \frac{m\vec{b} - n\vec{a}}{m-n}.$$

2 VECTORS IN THE PLANE   59

Demonstration 2 Express the points which divide the three medians of $\triangle ABC$ at a ratio of $2:1$ in terms of the position vectors of the three vertices. Prove that the three medians intersect at a single point.

[Proof] Take $\vec{a}$, $\vec{b}$, and $\vec{c}$ as the position vectors of vertices A, B, and C of $\triangle ABC$. Take $L(\vec{l})$ as the midpoint of side BC.

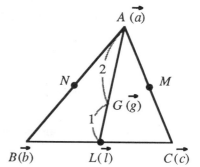

$$\vec{l} = \frac{\vec{b} + \vec{c}}{2} \qquad (1)$$

Take $G(\vec{g})$ as the point which divides median AL internally at a ratio of $2:1$.

$$\vec{g} = \frac{2\vec{l} + \vec{a}}{3} \qquad (2)$$

From (1) and (2),

$$\vec{g} = \frac{\vec{a} + \vec{b} + \vec{c}}{3}.$$

If we find the position vectors of the points which divide medians BM and CN at a ratio of $2:1$, we get the same result as above.

Therefore, the three medians of a triangle intersect at a single point.

Point G in Demonstration 2 is the centroid of $\triangle ABC$.

Problem 3 The line segment connecting the midpoints of two sides of a triangle is parallel to the third side, and it is half as long as the third side. Prove this statement using position vectors.

Coordinates and Position Vectors

Given a set of coordinate axes, the origin is usually taken as the standard initial point for position vectors.

Take (x, y) as the coordinates of point P, and take $\vec{p}$ as the position vector of P. Then

$$\vec{p} = \overrightarrow{OP} = x\vec{e}_1 + y\vec{e}_2.$$

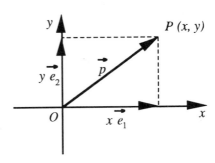

By component representation, $\vec{p} = (x, y)$.

Thus, if we take the origin as the initial point, the component representation of the position vector of P contains the coordinates of P.

 Problem 4   Prove that the coordinates of point C, which internally divides line segment AB connecting two points $A(x_1, y_1)$ and $B(x_2, y_2)$ at a ratio of $m:n$, are given by the following formulas, based on Demonstration 1.

$$x = \frac{mx_2 + nx_1}{m + n}, \quad y = \frac{my_2 + ny_1}{m + n}$$

② Straight Lines and Vectors

Straight Lines and Direction Vectors; Parametric Representation of Lines

Let's consider a straight line l through a fixed point $P_0(\vec{p}_0)$ and parallel to a given vector $\vec{u}$ not equal to $\vec{0}$.

Take P as a moving point on l, and then

$$\overrightarrow{P_0P} \parallel \vec{u}.$$

Therefore, we can write

$$\overrightarrow{P_0P} = t\vec{u}$$

for a real number t.

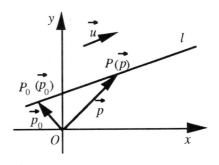

Take $\vec{p}$ as the position vector of moving point P, and then $\overrightarrow{P_0P} = \vec{p} - \vec{p}_0$.
Therefore,

$$\vec{p} - \vec{p}_0 = t\vec{u}.$$

Thus,

$$\vec{p} = \vec{p}_0 + t\vec{u}. \qquad (1)$$

Take (x_0, y_0) as the coordinates of fixed point P_0, take (x, y) as the coordinates of moving point P, and let $\vec{u} = (a, b)$. Then in (1),

$$\vec{p} = (x, y),\ \vec{p}_0 = (x_0, y_0),\ \vec{u} = (a, b).$$

Therefore,

$$(x, y) = (x_0, y_0) + t(a, b) = (x_0 + ta, y_0 + tb).$$

Thus, (1) can be rewritten in the following form:

$$\begin{cases} x = x_0 + ta \\ y = y_0 + tb \end{cases} \qquad (2)$$

(1) and (2) above are referred to as the **parametric representation** of straight line l, and t is the **parameter**. To be more exact, (1) is the parametric representation in the form of a vector, while (2) is the parametric representation by components. Moreover, $\vec{u} = (a, b)$ is called the **direction vector** of straight line l.

If parameter t ranges over the set of all real numbers, then in (1) or (2) point $P(\vec{p})$ or point $P(x, y)$ moves along all points on l.

Note: If $k \neq 0$, then $k\vec{u}$ is also the direction vector of a straight line.

Problem 1 Express the following straight lines in the form given in (2) above:

(1) a straight line through the point (2, -3) with a direction vector of (1, 2).

(2) a straight line through the point (4, 0) with a direction vector of (-3, 2).

For $a \neq 0$ and $b \neq 0$, eliminating t from (2) above gives us

$$\frac{x - x_0}{a} = \frac{y - y_0}{b}.$$

Therefore,

$$y - y_0 = \frac{b}{a}(x - x_0).$$

This equation is identical to the one we learned as the equation of a straight line through the point (x_0, y_0) with a slope of $\frac{b}{a}$.

Demonstration 1 The parametric representation of straight line l through two points $A(\vec{a})$ and $B(\vec{b})$ in the form of a vector can be expressed in the following form, with $P(\vec{p})$ as a point moving along l.

$$\vec{p} = \vec{a} + t(\vec{b} - \vec{a})$$

Prove this statement.

[Proof] Since l passes through the two points A and B, we can take

$$\vec{AB} = \vec{b} - \vec{a}$$

as the direction vector.
Line l also passes through point $A(\vec{a})$.

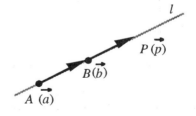

Therefore, the parametric representation of l in the form of a vector is

$$\vec{p} = \vec{a} + t(\vec{b} - \vec{a}).$$

Problem 2 Find the direction vector $\vec{AB}$ of a straight line that passes through the following pairs of points A and B. Then find the parametric representation of the straight lines in the form of a vector.

(1) $A(-3, 2)$, $B(4, 5)$ 　　　　　(2) $A(4, 0)$, $B(0, 3)$

(3) $A(4, -3)$, $B(-2, 5)$ 　　　　(4) $A(-\frac{3}{2}, -4)$, $B(2, -4)$

Demonstration 2 The necessary and sufficient condition such that point $C(\vec{c})$ lies on straight line AB, where $A(\vec{a})$ and $B(\vec{b})$ are two different points, is that there exist real numbers m and n which satisfy

$$\vec{c} = m\vec{a} + n\vec{b}, \quad m + n = 1.$$

Prove this statement.

[Proof] If $C(\vec{c})$ lies on straight line AB, we know from Demonstration 1 that there exists a real number t_0 satisfying

$$\vec{c} = \vec{a} + t_0(\vec{b} - \vec{a}). \tag{1}$$

Rewriting (1), we obtain

$$\vec{c} = (1 - t_0)\vec{a} + t_0\vec{b}. \tag{2}$$

Therefore, take $1 - t_0 = m$ and $t_0 = n$, and then the following equalities hold:

$$\vec{c} = m\vec{a} + n\vec{b}, \quad m + n = 1. \tag{3}$$

Conversely, if (3) holds, we can take $n = t_0$, and then (3) can be rewritten as (2) and as (1). Therefore, $C(\vec{c})$ lies on straight line AB.

Problem 3 In Demonstration 2, the necessary and sufficient condition such that point $C(\vec{c})$ lies on line segment AB is

$$\vec{c} = m\vec{a} + n\vec{b}, \quad m \geq 0, \ n \geq 0, \ m + n = 1.$$

Prove this statement.

Straight Lines and Normal Vectors

Let's find the equation of a straight line l through a fixed point $P_0(\vec{p}_0)$ and perpendicular to a given vector $\vec{n}$ not equal to $\vec{0}$.

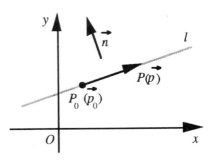

Take $P(\vec{p})$ as a point moving along l, and then

$$\overrightarrow{P_0P} = \vec{p} - \vec{p}_0, \quad \overrightarrow{P_0P} \perp \vec{n}.$$

Therefore,

$$\vec{n} \cdot (\vec{p} - \vec{p}_0) = 0. \qquad (1)$$

This equality expresses the condition that point $P(\vec{p})$ lies on line l.

In general, given figure F in a plane, an equality which expresses in terms of vectors the condition that point P lies in F is called a **vector equation**. Equation (1) above is the vector equation of line l.

Let us express all the vectors in (1) in terms of their components:

$$\vec{p} = (x, y), \quad \vec{p}_0 = (x_0, y_0), \quad \vec{n} = (a, b).$$

Then

$$\vec{p} - \vec{p}_0 = (x - x_0, y - y_0)$$

$$\vec{n} \cdot (\vec{p} - \vec{p}_0) = a(x - x_0) + b(y - y_0).$$

Therefore, (1) takes the form

$$a(x - x_0) + b(y - y_0) = 0.$$

The equation of a straight line through point $P_0(x_0, y_0)$ and perpendicular to vector $\vec{n} = (a, b)$ is

$$a(x - x_0) + b(y - y_0) = 0. \qquad (2)$$

Vector $\vec{n} = (a, b)$ is called the **normal vector** of this straight line.

Note: If $k \neq 0$, then $k\vec{n}$ is also a normal vector of this straight line.

In (2) above, if we set $c = -ax_0 - by_0$, then (2) can be rewritten as

$$ax + by + c = 0.$$

This is the form of the equation for a straight line which you learned in grade 10 mathematics.

Moreover, we can see from the above discussion that one normal vector of a straight line expressed by the equation

$$ax + by + c = 0$$

is (a, b).

Problem 4 Find the normal vector of the line $3x - 4y + 5 = 0$.

Problem 5 Find the equation of a straight line through the point $(5, -4)$ and perpendicular to vector $\vec{n} = (2, 3)$.

Demonstration 3 Take the straight line $ax + by + c = 0$ as l, and take point $P(x_1, y_1)$ as one point not on l. Take d as the distance between P and l, that is, the length of a perpendicular line from P to l. Prove that d has the following value using vectors.

$$d = \frac{|ax_1 + by_1 + c|}{\sqrt{a^2 + b^2}}$$

[Proof] Take (x_2, y_2) as the coordinates of point H, and then

$$\vec{PH} = (x_2 - x_1, y_2 - y_1).$$

If we take $\vec{n} = (a, b)$, then

$$\vec{PH} \parallel \vec{n}.$$

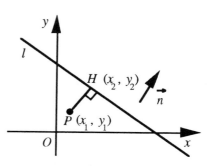

Therefore,

$$\overrightarrow{PH} \cdot \vec{n} = \pm |\overrightarrow{PH}| |\vec{n}|$$

$$|\overrightarrow{PH} \cdot \vec{n}| = |\overrightarrow{PH}| |\vec{n}| = d|\vec{n}|.$$

Thus,

$$d = \frac{|\overrightarrow{PH} \cdot \vec{n}|}{|\vec{n}|} = \frac{|a(x_2 - x_1) + b(y_2 - y_1)|}{\sqrt{a^2 + b^2}}.$$

Here point H lies on l, and then

$$ax_2 + by_2 + c = 0.$$

Therefore,

$$a(x_2 - x_1) + b(y_2 - y_1) = -ax_1 - by_1 - c.$$

Thus,

$$d = \frac{|ax_1 + by_1 + c|}{\sqrt{a^2 + b^2}}.$$

Problem 6 Find the length of a perpendicular line from point $P(2, -5)$ to the straight line $4x - 3y + 7 = 0$.

3 Circles and Vectors

Take point $P(\vec{p})$ as an arbitrary point on a circle with its center at point $C(\vec{c})$ and a radius of r. Then

$$|\overrightarrow{CP}| = r.$$

Thus,

$$|\vec{p} - \vec{c}| = r. \qquad (1)$$

This is the vector equation of a circle.

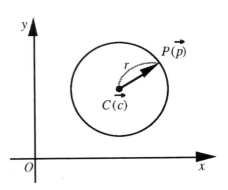

Squaring both sides of (1), we obtain

$$|\vec{p} - \vec{c}|^2 = r^2.$$

Let $\vec{p} = (x, y)$ and $\vec{c} = (x_0, y_0)$ in this formula, and we get

$$(x - x_0)^2 + (y - y_0)^2 = r^2.$$

This equation is identical to the equation of a circle which you learned in grade 10.

Problem 1 Prove that the vector equation of a circle in which the two points $A(\vec{a})$ and $B(\vec{b})$ are the end points of a diameter is given by

$$(\vec{p} - \vec{a}) \cdot (\vec{p} - \vec{b}) = 0$$

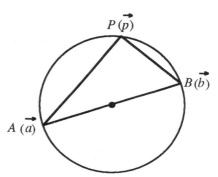

where $P(\vec{p})$ is an arbitrary point on the circle.

Problem 2 Take one point $P_0(\vec{p}_0)$ on a circle with center $C(\vec{c})$ and radius r. Show that the vector equation of a line tangent to this circle at point P_0 can be expressed as

$$(\vec{p} - \vec{c}) \cdot (\vec{p}_0 - \vec{c}) = r^2$$

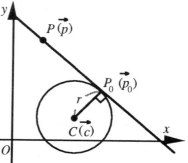

where $P(p)$ is an arbitrary point on the tangent line.

68 2 VECTORS IN THE PLANE

Applying Vectors to Figures

Demonstration 1 Take M as the midpoint of side BC of $\triangle ABC$, and then

$$AB^2 + AC^2 = 2(AM^2 + BM^2).$$

Prove this.

[**Proof**]   Take M as the origin as in the figure to the right, and take $\vec{a}$ and $\vec{b}$ as the position vectors of A and B. Then the position vector of C is $-\vec{b}$. Therefore,

$$\vec{BA} = \vec{a} - \vec{b}$$

$$\vec{CA} = \vec{a} + \vec{b}.$$

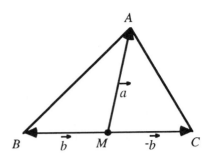

Thus,

$$AB^2 = |\vec{BA}|^2 = (\vec{a} - \vec{b})\cdot(\vec{a} - \vec{b}) = |\vec{a}|^2 - 2\vec{a}\cdot\vec{b} + |\vec{b}|^2$$

$$AC^2 = |\vec{CA}|^2 = (\vec{a} + \vec{b})\cdot(\vec{a} + \vec{b}) = |\vec{a}|^2 + 2\vec{a}\cdot\vec{b} + |\vec{b}|^2.$$

Therefore,

$$AB^2 + AC^2 = 2(|\vec{a}|^2 + |\vec{b}|^2) = 2(AM^2 + BM^2).$$

Problem 1 Take D as the point which divides side BC of $\triangle ABC$ internally at a ratio of $m:n$, and then

$$nAB^2 + mAC^2 = nBD^2 + mCD^2$$
$$+ (m+n)AD^2.$$

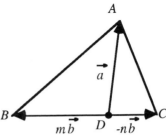

Prove this using vectors.

2 VECTORS IN THE PLANE   69

Demonstration 2   Prove that three perpendicular lines from the three vertices of △ABC to the opposite sides all intersect at one point.

[**Proof**]   It is sufficient to take H as the intersection of the perpendicular lines from vertices B and C to the opposite sides and prove that the perpendicular line from A to the opposite side passes through H.

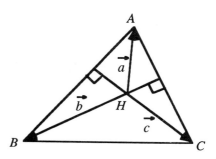

Take H as the origin and take $\vec{a}$, $\vec{b}$, and $\vec{c}$ as the position vectors of A, B, and C. Since $\vec{c}$ is perpendicular to $\overrightarrow{AB}$,

$$(\vec{b} - \vec{a}) \cdot \vec{c} = 0.$$

Thus,

$$\vec{b} \cdot \vec{c} = \vec{a} \cdot \vec{c}. \tag{1}$$

Since $\vec{b}$ is perpendicular to $\overrightarrow{AC}$,

$$(\vec{c} - \vec{a}) \cdot \vec{b} = 0.$$

Thus,

$$\vec{c} \cdot \vec{b} = \vec{a} \cdot \vec{b}. \tag{2}$$

The left sides of (1) and (2) are identical, and therefore,

$$\vec{a} \cdot \vec{c} = \vec{a} \cdot \vec{b}.$$

Thus,

$$\vec{a} \cdot (\vec{c} - \vec{b}) = 0.$$

Therefore, $\vec{a}$ is perpendicular to $\overrightarrow{BC}$, and so the line from A perpendicular to BC passes through H.

Point H in Demonstration 2 is the orthocenter of △ABC.

70   2 VECTORS IN THE PLANE

Problem 2   Take H as the orthocenter of $\triangle ABC$, and take point P satisfying

$$\overrightarrow{HP} = \frac{1}{2}(\overrightarrow{HA} + \overrightarrow{HB} + \overrightarrow{HC}).$$

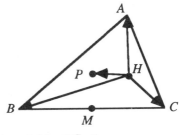

Prove that:

(1) If we take M as the midpoint of side BC, then

$$\overrightarrow{HA} = 2\overrightarrow{MP}.$$

(2) P is the circumcenter of $\triangle ABC$.

  ## Force, Velocity, and Vectors

Force Vectors

Vectors are used to express the force applied at a point. The direction of the vector is the direction of the force, and the magnitude is proportional to the magnitude of the force. This is called a **force vector**.

For the case when two forces are applied at single point, we know the following "parallelogram rule of forces" from physics.

When two forces expressed by vectors $\vec{f}$ and $\vec{g}$ are applied to a single point A, the combined force is expressed as

$$\vec{f} + \vec{g}.$$

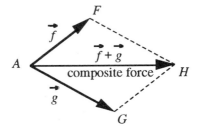

Therefore, the effect of two forces $\vec{f}$ and $\vec{g}$ applied at A is identical to the effect of one force $\vec{f} + \vec{g}$ applied at A.

If two forces f_1 and f_2 applied at a single point are balanced, the vector representing the combined forces is a zero vector. That is, the following equality holds:

$$f_1 + f_2 = 0.$$

We can deal with three or more forces analogously.

2 VECTORS IN THE PLANE

Problem 1 Three forces $\vec{f}$, $\vec{g}$, and $\vec{h}$ are applied at a single point O, and they are balanced. Given $\vec{f}$ and $\vec{g}$ as in the figure to the right, sketch $\vec{h}$ in the same figure.

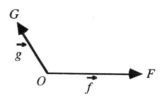

Problem 2 As in the figure to the right, we hang an object of 30 kg at point C on a rope that is fixed to two points A and B by its end points. Find the forces applied at points A and B.

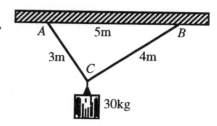

Velocity Vectors

For a moving point P, one vector $\vec{v}$, called a **velocity vector**, is defined for each point in time.

The direction of $\vec{v}$ is the direction in which P moves at time t, and the magnitude of $\vec{v}$ is the velocity of point P at time t, that is, the distance it moves in a unit of time.

In other words, the velocity vector of point P at time t is the vector expressing the movement of P per unit of time at t.

Problem 3 A river flows at 2 m/second from west to east. We row a boat at 2 m/second from south to north perpendicular to the river current. What is the actual velocity in meters per second and the direction of the boat?

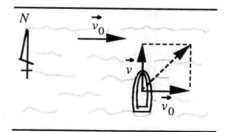

Exercises

1. Take $\vec{a}$ and $\vec{b}$ as the position vectors of two points A and B. Express the position vectors of the following points in terms of $\vec{a}$ and $\vec{b}$.

    (1) The point which divides line segment AB internally at a ratio of $3:2$.

    (2) The point which divides line segment AB externally at a ratio of $1:2$.

    (3) The point symmetric to A with respect to B.

2. $\vec{a}$ and $\vec{b}$ are non-parallel vectors, and the position vectors $\vec{p}$, $\vec{q}$, and $\vec{r}$ of three points P, Q, and R are expressed as follows:

    $$\vec{p} = 2\vec{a} + 2\vec{b}, \quad \vec{q} = -6\vec{a} + 6\vec{b}, \quad \vec{r} = 6\vec{a}$$

    (1) Express $\vec{PQ}$ and $\vec{PR}$ in terms of $\vec{a}$ and $\vec{b}$.

    (2) What is the relation among the three points P, Q, and R?

3. Take $\vec{a}$ and $\vec{b}$ as the position vectors of two points A and B, where point O is the shared initial point. Find the vector equations of the bisectors of $\angle AOB$ for:

    (1) $|\vec{a}| = |\vec{b}| = 1$ (2) $|\vec{a}| = 2, |\vec{b}| = 3$

4. l_1 is a straight line passing through the point $(1, 1)$ with a direction vector $\vec{u}_1 = (1, 2)$; l_2 is a straight line passing through the point $(1, 5)$ with a direction vector $\vec{u}_2 = (3, -4)$.

    (1) Find the parametric representation, taking the components of l_1 and l_2 as the parameters s and t.

    (2) Find the coordinates of the point at which l_1 intersects l_2.

5. What kind of quadrilateral is $ABCD$ if the following relations hold?

    (1) $\vec{AC} + \vec{BD} = 2\vec{AD}$

    (2) $\vec{AD} = \vec{AC} - \vec{AB}$ and $(\vec{AB} - \vec{AD}) \cdot (\vec{AD} - \vec{CD}) = 0$

Chapter Exercises

A

1. Find the coordinates of points P and Q which satisfy the following conditions, given the two points $A(-3, 4)$ and $B(2, -1)$. O is the origin.

    (1) $\vec{PO} = \vec{AB}$
    (2) $\vec{AQ} = \frac{1}{2}\vec{AB}$

2. Prove the following statements, given $\vec{OA} = 2\vec{a}$, $\vec{OB} = 3\vec{b}$, $\vec{OC} = 6\vec{a} - 6\vec{b}$, and $\vec{OD} = 6\vec{b} - 4\vec{a}$.

    (1) The three points A, B, and C all lie on a single line.
    (2) $AB \parallel OD$

3. Find the following inner products, given $|\vec{a}| = \sqrt{3}$, $|\vec{b}| = 2$, and $|\vec{a} + \vec{b}| = 1$.

    (1) $\vec{a} \cdot \vec{b}$
    (2) $(\vec{a} - \vec{b}) \cdot (\vec{a} + 2\vec{b})$

4. Given $\vec{OP} = (1, 1)$ and $\vec{OQ} = (1 - \sqrt{3}, 1 + \sqrt{3})$.

    (1) Find the angle formed by $\vec{OP}$ and $\vec{OQ}$.
    (2) Find the area of $\triangle OPQ$.

5. Find the normal vectors of two straight lines $\sqrt{3}x + y - 1 = 0$ and $x + \sqrt{3}y + 2 = 0$, and find the angle formed by these two lines.

6. For the arbitrary vectors $\vec{p}$ and $\vec{q}$, $|\vec{p} \cdot \vec{q}| \le |\vec{p}||\vec{q}|$ holds. Use this fact to prove the following inequality, assuming that $\vec{p} = (a, b)$ and $\vec{q} = (x, y)$.

    $$(ax + by)^2 \le (a^2 + b^2)(x^2 + y^2)$$

7. Point P moves in a plane with a velocity vector $\vec{v} = (2, 5)$. When $t = 0$, P is located at point $A(-6, -2)$; the unit of time is one second.

    (1) Find the position vector $\vec{p}$ of P after t seconds.
    (2) When does P approach closest to the point $(0, 2)$?

74   2 VECTORS IN THE PLANE

B

1. In parallelogram $ABCD$, take E as the point which divides side AB internally at a ratio of $2:1$, and take F as the point which divides diagonal BD internally at a ratio of $1:3$.

   (1) Let $\vec{BA} = \vec{a}$ and $\vec{BC} = \vec{b}$, and express $\vec{CE}$ and $\vec{CF}$ in terms of $\vec{a}$ and $\vec{b}$.

   (2) Prove that the three points C, E, and F all lie on one straight line.

2. In $\triangle ABC$ with three vertices $A(\vec{a})$, $B(\vec{b})$, and $C(\vec{c})$, take $P(\vec{p})$ as the point which divides AB internally at a ratio of $1:2$, $Q(\vec{q})$ as the midpoint of AC, and $R(\vec{r})$ as the point which divides BC internally at a ratio of $2:1$. Then derive the formula $\vec{q} = \frac{3}{4}\vec{p} + \frac{1}{4}\vec{r}$, and prove that the three points P, Q, and R all lie on one straight line.

3. Take S as the area of a parallelogram in which two non-parallel vectors $\vec{a}$ and $\vec{b}$ form two sides. Then prove that $S^2 = |\vec{a}|^2|\vec{b}|^2 - (\vec{a} \cdot \vec{b})^2$. Also prove that $S = |a_1 b_2 - a_2 b_1|$, taking $\vec{a} = (a_1, a_2)$ and $\vec{b} = (b_1, b_2)$.

4. Given that $\vec{a} \neq 0$, $|\vec{b}| = 2|\vec{a}|$, and $\vec{a} + \vec{b}$ and $5\vec{a} - 2\vec{b}$ are perpendicular, find the angle formed by the two vectors $\vec{a}$ and $\vec{b}$.

5. The origin O and the two points $A(\vec{a})$ and $B(\vec{b})$ do not lie on a single straight line. Prove that the position vector $\vec{c}$ of point C, which lies inside $\triangle OAB$, can be expressed as

   $$\vec{c} = m\vec{a} + n\vec{b}, \quad m > 0, \; n > 0, \; m + n < 1$$

6. In the figure to the right, quadrilaterals $ABDE$ and $ACFG$ are both squares. Taking M as the midpoint of line segment EG, prove that line MA is perpendicular to line BC by using vectors.

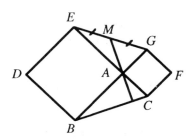

CHAPTER 3

MATRICES

SECTION 1. MATRICES
SECTION 2. LINEAR TRANSFORMATIONS

It was Arthur Cayley (1821–1895) who first advanced the concept of a matrix. Historically, the concept of a determinant (which is called "the expression of a matrix" in Japanese) preceded the matrix; it originated in the general solution of simultaneous linear equations. The origin of the concept of a determinant is noted in a 1678 letter of Leibniz, while in our country Takakazu Seki made great achievements in the 1680s.

In the second half of the 19th century, Cayley, James Sylvester, and Ferdinand Frobenius studied the theory of matrices and determinants in great depth. Along with advances in linear algebra, matrices were recognized as closely related to linear transformations or linear mappings, and they have become one of the central concepts of the theory of linear algebra.

Moreover, in this century the theory of matrices has become indispensable in various fields such as physics, engineering, and economics.

 MATRICES

The Meaning of a Matrix

The component representation of a vector which you learned in Chapter 2 is a horizontal arrangement of two numbers. In this chapter, let's consider arrangements of several numbers in the form of a rectangle.

An arrangement of several numbers in the form of a rectangle, for example,

$$\begin{pmatrix} 1 & 2 \\ 0 & 5 \end{pmatrix} \qquad (1)$$

$$\begin{pmatrix} 2 & 1 & 5 \\ 3 & -4 & 6 \end{pmatrix} \qquad (2)$$

is called a **matrix**. Matrices are usually written with parentheses at the sides, as above.

In a matrix the horizontal arrangements of numbers, starting from the top, are referred to as the **first row, second row,...**, and the vertical arrangements of numbers, starting from the left, are referred to as the **first column, second column, ...**

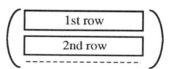

The number at the intersection of the ith row and jth column is called the (i, j) **component**.

A matrix consisting of m rows and n columns is referred to as a matrix with m **rows and** n **columns** or $m \times n$ **matrix**. As a special case, an $n \times n$ matrix is called a **square matrix of dimension** n.

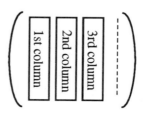

A square matrix of dimension 1 is nothing but one number.

Example 1 Matrix (1) at the top of this page is a square matrix of dimension 2, and matrix (2) is a 2 × 3 matrix.

Example 2 In matrix (2) on the preceding page,
the first row is (2 1 5)

the second column is $\begin{pmatrix} 1 \\ -4 \end{pmatrix}$

the (1, 2) component is 1.

Problem 1 Write out the first row, the second row, the first column, and the second column of matrix (1) on the preceding page. Then give the (1, 1), (1, 2), (2, 1), and (2, 2) components.

A matrix consisting of one row is called a **row vector**, and a matrix consisting of one column is called a **column vector**. For example, a 1 x 4 matrix is called a row vector of **dimension 4**, and a 3 x 1 matrix is called a column vector of **dimension 3**.

Row vectors and column vectors, taken together, are called **numerical vectors**. They are sometimes simply called **vectors**.

Example 3 $(2 \ -5 \ \sqrt{10})$ is a row vector of dimension 3.

$\begin{pmatrix} 10 \\ 20 \end{pmatrix}$ is a column vector of dimension 2.

Matrices are expressed by capital letters, such as A or B, and each component is expressed by a small letter or a small letter with a subscript. For example,

$$A = \begin{pmatrix} a & b \\ c & d \end{pmatrix} \quad \text{or} \quad A = \begin{pmatrix} a_{11} & a_{12} \\ a_{21} & a_{22} \end{pmatrix}.$$

Here, a_{ij} means the (i, j) component of matrix A.

Identical Matrices

If two matrices A and B have the same form, and all the corresponding components are the same, they are said to be **identical**, and are represented as $A = B$.

Example 4 $\begin{pmatrix} a & 0 \\ b & 1 \end{pmatrix} = \begin{pmatrix} -1 & c \\ 2 & d \end{pmatrix}$

is equivalent to setting $a = -1, b = 2, c = 0,$ and $d = 1$.

3 MATRICES

Problem 2 Find the value of $x, y, z,$ and u, if

$$\begin{pmatrix} x & 2y \\ z & -6 \end{pmatrix} = \begin{pmatrix} -2u & u \\ 4 & 3x \end{pmatrix}.$$

  Addition, Subtraction, and Multiplication Involving Matrices

Addition and Subtraction of Matrices

Example 1 The following table shows the quantity of products manufactured at a certain factory in April and May. Three different kinds of products are manufactured: $P, Q,$ and R; and there are two factories (the quantities are given in tons).

April	P	Q	R
Factory 1	5.7	2.4	1.7
Factory 2	8.5	4.3	2.8

May	P	Q	R
Factory 1	6.0	2.2	2.0
Factory 2	8.2	4.5	3.0

If we compile a new table, based on the one above, to show the total quantity produced during April and May, we obtain the following table.

April and May	P	Q	R
Factory 1	11.7	4.6	3.7
Factory 2	16.7	8.8	5.8

Therefore, by adding together the corresponding components of the two matrices representing the quantities produced in April and May

$$\begin{pmatrix} 5.7 & 2.4 & 1.7 \\ 8.5 & 4.3 & 2.8 \end{pmatrix} \quad \begin{pmatrix} 6.0 & 2.2 & 2.0 \\ 8.2 & 4.5 & 3.0 \end{pmatrix}$$

we get a new matrix representing the total production for April and May.

$$\begin{pmatrix} 5.7 + 6.0 & 2.4 + 2.2 & 1.7 + 2.0 \\ 8.5 + 8.2 & 4.3 + 4.5 & 2.8 + 3.0 \end{pmatrix} = \begin{pmatrix} 11.7 & 4.6 & 3.7 \\ 16.7 & 8.8 & 5.8 \end{pmatrix}$$

Based on the idea illustrated in Example 1, we can define the **sum** $A + B$ of two matrices of the same form in the following way:

The Sum of Two Matrices

Given $A = \begin{pmatrix} a_{11} & a_{12} & a_{13} \\ a_{21} & a_{22} & a_{23} \end{pmatrix}$, $B = \begin{pmatrix} b_{11} & b_{12} & b_{13} \\ b_{21} & b_{22} & b_{23} \end{pmatrix}$

$A + B = \begin{pmatrix} a_{11} + b_{11} & a_{12} + b_{12} & a_{13} + b_{13} \\ a_{21} + b_{21} & a_{22} + b_{22} & a_{23} + b_{23} \end{pmatrix}$

The sum of two matrices of any form can be defined analogously.

Note: The sum $A + B$ of matrices A and B is defined only when A and B have the same form. For example, we cannot speak of the sum of a 2 × 2 matrix and a 2 × 3 matrix.

Problem 1 Find the sum of the following matrices:

(1) $\begin{pmatrix} 2 & -3 & 4 \\ 1 & 0 & -1 \end{pmatrix} + \begin{pmatrix} 5 & 1 & 0 \\ 3 & 2 & -4 \end{pmatrix}$

(2) $\begin{pmatrix} 1 & -2 \\ -2 & 3 \end{pmatrix} + \begin{pmatrix} 2 & 1 \\ 0 & -1 \end{pmatrix}$

(3) $\begin{pmatrix} 2 \\ 1 \\ 3 \end{pmatrix} + \begin{pmatrix} -4 \\ 0 \\ 5 \end{pmatrix}$

(4) $\begin{pmatrix} -7 & 5 & -1 & 2 \end{pmatrix} + \begin{pmatrix} 2 & -3 & -5 & 6 \end{pmatrix}$

When we add matrices, the following laws hold just as when we add vectors.

---

(1) $A + B = B + A$ Commutative law

(2) $(A + B) + C = A + (B + C)$ Associative law

---

3 MATRICES

A matrix in which all the components are 0 is called a **zero matrix**. For example,

$$\begin{pmatrix} 0 & 0 \\ 0 & 0 \end{pmatrix} \quad \begin{pmatrix} 0 & 0 & 0 \\ 0 & 0 & 0 \end{pmatrix} \quad \begin{pmatrix} 0 \\ 0 \end{pmatrix} \quad \begin{pmatrix} 0 & 0 \end{pmatrix}$$

are all zero matrices. These are not identical matrices, but if there is no chance of confusion, we can designate all of them by the same letter O.

Given any matrix A and a zero matrix of the same form as A, the following equality holds:

$$A + O = O + A = A.$$

Given matrix A, we designate the matrix whose components are given by changing the sign of each component of A as $-A$.

Example 2 If $A = \begin{pmatrix} 2 & -3 & 4 \\ 1 & 0 & -1 \end{pmatrix}$, then $-A = \begin{pmatrix} -2 & 3 & -4 \\ -1 & 0 & 1 \end{pmatrix}$.

Given matrices A and B of the same form, $A + (-B)$ is designated as

$$A - B$$

and is called the **difference** of A and B. It is nothing other than the matrix X which satisfies $B + X = A$.

Problem 2 Find the differences of the following matrices:

(1) $\begin{pmatrix} 1 & -4 \\ 2 & 3 \end{pmatrix} - \begin{pmatrix} -1 & 5 \\ 6 & -2 \end{pmatrix}$

(2) $\begin{pmatrix} 2 & 0 & -1 \\ -3 & -2 & 0 \end{pmatrix} - \begin{pmatrix} -1 & 3 & 4 \\ 1 & 2 & -5 \end{pmatrix}$

Problem 3 Find the matrix X satisfying $B + X = A$, if $A = \begin{pmatrix} 2 & -1 \\ 3 & 4 \end{pmatrix}$ and $B = \begin{pmatrix} -1 & 0 \\ 5 & 2 \end{pmatrix}$.

Multiplying a Matrix by a Real Number

Example 3 The factory referred to on page 78 institutes a plan to increase the quantity of each product manufactured in May by 10% in June. In this case, the matrix which represents the production plan for June is created by multiplying each component of the matrix for May

$$\begin{pmatrix} 6.0 & 2.2 & 2.0 \\ 8.2 & 4.5 & 3.0 \end{pmatrix}$$

by 1.1, so that we obtain

$$\begin{pmatrix} 6.6 & 2.42 & 2.2 \\ 9.02 & 4.95 & 3.3 \end{pmatrix}.$$

In general, given a real number k, the product kA of matrix A and k is defined in the following way:

Multiplying a Matrix by a Real Number

Given $A = \begin{pmatrix} a_{11} & a_{12} & a_{13} \\ a_{21} & a_{22} & a_{23} \end{pmatrix}$, $kA = \begin{pmatrix} ka_{11} & ka_{12} & ka_{13} \\ ka_{21} & ka_{22} & ka_{23} \end{pmatrix}.$

Problem 4 Find $3A$ and $\frac{1}{2}A$, given $A = \begin{pmatrix} 2 & 3 & 4 \\ 1 & 0 & -1 \end{pmatrix}.$

From the definition of a matrix multiplied by a real number, we have the following special cases:

$$1A = A, \quad (-1)A = -A$$

$$0A = O, \quad kO = O.$$

The following laws also hold:

---
(1) $(kl)A = k(lA)$ Associative law

(2) $(k + l)A = kA + lA$
(3) $k(A + B) = kA + kB$ } Distributive law

---

3 MATRICES

Problem 5 Check that $k(A + B) = kA + kB$, given $A = \begin{pmatrix} a_{11} & a_{12} \\ a_{21} & a_{22} \end{pmatrix}$ and $B = \begin{pmatrix} b_{11} & b_{12} \\ b_{21} & b_{22} \end{pmatrix}$.

Problem 6 Find the following matrices, given $A = \begin{pmatrix} 2 & 0 \\ -1 & 4 \end{pmatrix}$, $B = \begin{pmatrix} 1 & -7 \\ 3 & 0 \end{pmatrix}$ and $C = \begin{pmatrix} 0 & 5 \\ 8 & -2 \end{pmatrix}$.

(1) $3A - B + 2C$

(2) $2(A - B + 2C) + 3B - C$

Demonstration Find matrix X which satisfies the following equality for

$$A = \begin{pmatrix} 2 & -1 \\ 1 & -6 \end{pmatrix} \text{ and } B = \begin{pmatrix} 5 & -7 \\ 4 & 3 \end{pmatrix}:$$

$$A - X = 2(X - B)$$

[Solution] Eliminating the parentheses, transposing, and rearranging this equality, we obtain

$$-3X = -A - 2B.$$

Therefore,

$$X = \frac{1}{3}(A + 2B)$$

$$= \frac{1}{3}\left\{\begin{pmatrix} 2 & -1 \\ 1 & -6 \end{pmatrix} + 2\begin{pmatrix} 5 & -7 \\ 4 & 3 \end{pmatrix}\right\} = \begin{pmatrix} 4 & -5 \\ 3 & 0 \end{pmatrix}.$$

Problem 7 Given matrices A and B in the above Demonstration, find matrix X satisfying the following equalities:

(1) $X + A = 2(2B - X)$

(2) $3A + \frac{1}{2}X = X + 4A - \frac{1}{2}B$

3 Multiplication of Matrices

So far we have considered the sum and difference of two matrices, as well as multiplication of a matrix by a real number. Now let's consider how to multiply two matrices.

Example 1 The figure to the right shows the unit price of a notebook and a pencil, and the quantities of them purchased.

The total purchase price can be calculated in the following way:

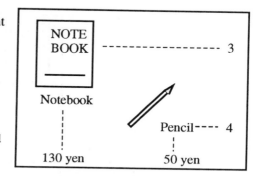

130 × 3 + 50 × 4 = 590 (yen)

The calculation in Example 1 requires that we find the products of the corresponding components of the row vector (130 50) and the column vector $\begin{pmatrix} 3 \\ 4 \end{pmatrix}$, 130 × 3 and 50 × 4, and then find their sum. We can demonstrate this calculation in the figure to the right.

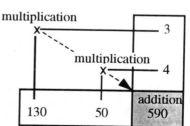

Based on this idea, we can define the product of a row vector and a column vector as:

$$(a_1 \ a_2) \begin{pmatrix} b_1 \\ b_2 \end{pmatrix} = a_1 b_1 + a_2 b_2 .$$

Example 2 $(2 \ 5) \begin{pmatrix} -1 \\ 3 \end{pmatrix} = 2 \times (-1) + 5 \times 3 = 13$

Problem 1 Perform the following calculations:

(1) $(-1 \ \ 3) \begin{pmatrix} -2 \\ -1 \end{pmatrix}$ (2) $(\sin \theta \ \ \cos \theta) \begin{pmatrix} \sin \theta \\ \cos \theta \end{pmatrix}$

3 MATRICES

The product of a row vector and a column vector of a dimension other than 2 can be defined analogously, as long as each matrix is of the same dimension.

Example 3 The figure below shows the unit price of the goods in two shops P and Q, and the quantities purchased at the same time by a boy and a girl.

	Notebook	Pencil	Boy	Girl
			3	5
			4	2
Shop P	130 yen	50 yen	590 yen	750 yen
Shop Q	140 yen	45 yen	600 yen	790 yen

The matrix inside the double frame

$$\begin{pmatrix} 590 & 750 \\ 600 & 790 \end{pmatrix}$$

shows the total amount that would be spent by the boy and the girl at shop P and Q. The (1, 1) component of this matrix represents the case when the boy makes his purchases at shop P, and is the same as Example 1. The (1, 2) component is the case when the girl makes her purchases at shop P, and it is calculated as

$$\begin{pmatrix} 130 & 50 \end{pmatrix} \begin{pmatrix} 5 \\ 2 \end{pmatrix} = 130 \times 5 + 50 \times 2 = 750 \quad \text{(yen)}.$$

The other components can be calculated analogously.

Based on the idea illustrated in Example 3, the **product** AB of two 2 × 2 matrices A and B can be defined in the following way:

The Product of Matrices

Given $A = \begin{pmatrix} a_{11} & a_{12} \\ a_{21} & a_{22} \end{pmatrix}$, and $B = \begin{pmatrix} b_{11} & b_{12} \\ b_{21} & b_{22} \end{pmatrix}$.

$$AB = \begin{pmatrix} a_{11}b_{11} + a_{12}b_{21} & a_{11}b_{12} + a_{12}b_{22} \\ a_{21}b_{11} + a_{22}b_{21} & a_{21}b_{12} + a_{22}b_{22} \end{pmatrix}$$

We can generalize this idea so that as long as the number of columns in matrix A is equal to the number of rows in matrix B, their product can be defined.

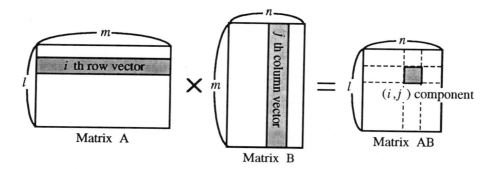

The (i, j) Component of a Product

The product AB of matrices A and B can be defined only when the number of columns of A is equal to the number of rows of B.

The (i, j) component of AB is the product of the ith row vector of A and the jth column vector of B.

3 MATRICES

Demonstration Calculate the product of the following pairs of matrices.

(1) $\begin{pmatrix} 1 & 2 \\ 3 & 4 \end{pmatrix} \begin{pmatrix} 6 & 4 \\ 5 & 3 \end{pmatrix}$ 　　　(2) $\begin{pmatrix} 2 & 1 \\ 1 & 3 \end{pmatrix} \begin{pmatrix} 4 \\ 5 \end{pmatrix}$

(3) $\begin{pmatrix} 1 \\ -2 \end{pmatrix} \begin{pmatrix} 3 & 1 \end{pmatrix}$

[Solution]

(1) $\begin{pmatrix} 1 & 2 \\ 3 & 4 \end{pmatrix} \begin{pmatrix} 6 & 4 \\ 5 & 3 \end{pmatrix} = \begin{pmatrix} 1\cdot 6 + 2\cdot 5 & 1\cdot 4 + 2\cdot 3 \\ 3\cdot 6 + 4\cdot 5 & 3\cdot 4 + 4\cdot 3 \end{pmatrix}$

$= \begin{pmatrix} 16 & 10 \\ 38 & 24 \end{pmatrix}$

(2) $\begin{pmatrix} 2 & 1 \\ 1 & 3 \end{pmatrix} \begin{pmatrix} 4 \\ 5 \end{pmatrix} = \begin{pmatrix} 2\cdot 4 + 1\cdot 5 \\ 1\cdot 4 + 3\cdot 5 \end{pmatrix} = \begin{pmatrix} 13 \\ 19 \end{pmatrix}$

(3) $\begin{pmatrix} 1 \\ -2 \end{pmatrix} \begin{pmatrix} 3 & 1 \end{pmatrix} = \begin{pmatrix} 1\cdot 3 & 1\cdot 1 \\ -2\cdot 3 & -2\cdot 1 \end{pmatrix} = \begin{pmatrix} 3 & 1 \\ -6 & -2 \end{pmatrix}$

Problem 2 Calculate the product of the following pairs of matrices:

(1) $\begin{pmatrix} 1 & -2 \\ 3 & 4 \end{pmatrix} \begin{pmatrix} 3 & 4 \\ 1 & 2 \end{pmatrix}$ 　　　(2) $\begin{pmatrix} 7 & 5 \\ 4 & 2 \end{pmatrix} \begin{pmatrix} -1 & 2 \\ 0 & 3 \end{pmatrix}$

(3) $\begin{pmatrix} -2 & 3 \\ 5 & -1 \end{pmatrix} \begin{pmatrix} 1 & 0 \\ 1 & -1 \end{pmatrix}$ 　　　(4) $\begin{pmatrix} 2 & 3 \\ -3 & 1 \end{pmatrix} \begin{pmatrix} 8 \\ -5 \end{pmatrix}$

(5) $\begin{pmatrix} -1 & 1 \end{pmatrix} \begin{pmatrix} 4 & -3 \\ -5 & 2 \end{pmatrix}$ 　　　(6) $\begin{pmatrix} 2 \\ 3 \end{pmatrix} \begin{pmatrix} -4 & 5 \end{pmatrix}$

Properties of Multiplication

We defined the multiplication of matrices in the preceding section; in this section, let's examine several properties of multiplication of 2 × 2 matrices.

Example 1 Let's find the product of matrices AB and BA for cases (1) and (2).

(1) Given $A = \begin{pmatrix} 1 & 0 \\ 0 & 0 \end{pmatrix}$, $B = \begin{pmatrix} -1 & 0 \\ 0 & 1 \end{pmatrix}$

$$AB = \begin{pmatrix} 1 & 0 \\ 0 & 0 \end{pmatrix}\begin{pmatrix} -1 & 0 \\ 0 & 1 \end{pmatrix} = \begin{pmatrix} -1 & 0 \\ 0 & 0 \end{pmatrix}$$

$$BA = \begin{pmatrix} -1 & 0 \\ 0 & 1 \end{pmatrix}\begin{pmatrix} 1 & 0 \\ 0 & 0 \end{pmatrix} = \begin{pmatrix} -1 & 0 \\ 0 & 0 \end{pmatrix}$$

(2) Given $A = \begin{pmatrix} 1 & 0 \\ 0 & 0 \end{pmatrix}$, $B = \begin{pmatrix} 0 & 1 \\ 1 & 0 \end{pmatrix}$

$$AB = \begin{pmatrix} 1 & 0 \\ 0 & 0 \end{pmatrix}\begin{pmatrix} 0 & 1 \\ 1 & 0 \end{pmatrix} = \begin{pmatrix} 0 & 1 \\ 0 & 0 \end{pmatrix}$$

$$BA = \begin{pmatrix} 0 & 1 \\ 1 & 0 \end{pmatrix}\begin{pmatrix} 1 & 0 \\ 0 & 0 \end{pmatrix} = \begin{pmatrix} 0 & 0 \\ 1 & 0 \end{pmatrix}$$

In part (1) of Example 1, $AB = BA$, but in (2), $AB \neq BA$. Thus, the commutative law,

$$AB = BA$$

does not hold generally for the product of two matrices. However, the following laws do hold for any real number k.

$k(AB) = (kA)B = A(kB)$

$(AB)C = A(BC)$ Associative law

$\left.\begin{aligned} A(B + C) &= AB + AC \\ (A + B)C &= AC + BC \end{aligned}\right\}$ Distributive law

---

Example 2 Let's check that the associative law holds for $A = \begin{pmatrix} 2 & 1 \\ 1 & 3 \end{pmatrix}$, $B = \begin{pmatrix} 3 & 4 \\ -1 & 2 \end{pmatrix}$, and $C = \begin{pmatrix} 1 & 2 \\ -2 & -1 \end{pmatrix}$.

$$AB = \begin{pmatrix} 2 & 1 \\ 1 & 3 \end{pmatrix}\begin{pmatrix} 3 & 4 \\ -1 & 2 \end{pmatrix} = \begin{pmatrix} 5 & 10 \\ 0 & 10 \end{pmatrix}$$

$$(AB)C = \begin{pmatrix} 5 & 10 \\ 0 & 10 \end{pmatrix}\begin{pmatrix} 1 & 2 \\ -2 & -1 \end{pmatrix} = \begin{pmatrix} -15 & 0 \\ -20 & -10 \end{pmatrix}$$

$$BC = \begin{pmatrix} 3 & 4 \\ -1 & 2 \end{pmatrix}\begin{pmatrix} 1 & 2 \\ -2 & -1 \end{pmatrix} = \begin{pmatrix} -5 & 2 \\ -5 & -4 \end{pmatrix}$$

$$A(BC) = \begin{pmatrix} 2 & 1 \\ 1 & 3 \end{pmatrix}\begin{pmatrix} -5 & 2 \\ -5 & -4 \end{pmatrix} = \begin{pmatrix} -15 & 0 \\ -20 & -10 \end{pmatrix}$$

Therefore, $(AB)C = A(BC)$.

Problem 1 Given matrices A, B, and C from Example 2, check that

$$A(B + C) = AB + AC \text{ and } (A + B)C = AC + BC$$

hold.

Since $(AB)C = A(BC)$, from the associative law, the product of three matrices can be written as ABC by eliminating the parentheses. The product of four or more matrices can be considered analogously.

Given a square matrix A, we can write AA, AAA, ... as $A^2, A^3, \ldots$.

Problem 2 Calculate the following expressions, given $A = \begin{pmatrix} -2 & 3 \\ 1 & 0 \end{pmatrix}$, $B = \begin{pmatrix} 4 & 3 \\ 2 & -1 \end{pmatrix}$, and $C = \begin{pmatrix} -1 & 2 \\ 3 & 4 \end{pmatrix}$.

(1) $AB - AC$ (2) $ABAB$ (3) A^4

Unit Matrix and Zero Matrix

The matrix

$$\begin{pmatrix} 1 & 0 \\ 0 & 1 \end{pmatrix}$$

is called the 2 × 2 **unit matrix**, and is designated as E.

For any 2 × 2 matrix A,

$$AE = EA = A$$

holds.

Problem 3 Check that $AE = EA = A$, taking $A = \begin{pmatrix} a & b \\ c & d \end{pmatrix}$.

For any 2 × 2 matrix A,

$$AO = OA = O$$

holds with the same form as a zero matrix.

Problem 4 Check the above statement.

When we perform operations on numbers,

if $a \neq 0$ and $b \neq 0$, then $ab \neq 0$.

However, when we perform operations on matrices, even if $A \neq O$ and $B \neq O$, it is possible that $AB = O$.

Example 3 Taking $A = \begin{pmatrix} -1 & 2 \\ 3 & -6 \end{pmatrix}$ and $B = \begin{pmatrix} 4 & 2 \\ 2 & 1 \end{pmatrix}$, then

$$AB = \begin{pmatrix} -1 & 2 \\ 3 & -6 \end{pmatrix}\begin{pmatrix} 4 & 2 \\ 2 & 1 \end{pmatrix} = \begin{pmatrix} -4+4 & -2+2 \\ 12-12 & 6-6 \end{pmatrix}$$

$$= \begin{pmatrix} 0 & 0 \\ 0 & 0 \end{pmatrix} = O.$$

3 MATRICES

The non-O matrices A and B which satisfy $AB = O$ are called **zero factors**. For example, A and B in Example 3 are zero factors.

Since there exist zero factors when matrices are multiplied, we cannot generally maintain the property that

$$\text{if } AB = O, \text{ then } A = O \text{ or } B = O.$$

Problem 5 Define the values of x and y such that $A = \begin{pmatrix} 1 & 1 \\ x & y \end{pmatrix}$ satisfies $A^2 = O$.

Problem 6 Show that the following property does not hold generally, using

$$A = \begin{pmatrix} -2 & -4 \\ 1 & 2 \end{pmatrix}, B = \begin{pmatrix} 3 & 4 \\ 7 & -1 \end{pmatrix}, \text{ and } C = \begin{pmatrix} 5 & -2 \\ 6 & 2 \end{pmatrix}:$$

"if $AB = AC$ and $A \neq O$, then $B = C$".

Demonstration For $A = \begin{pmatrix} a & b \\ c & d \end{pmatrix}$, prove the following equality:

$$A^2 - (a+d)A + (ad - bc)E = O$$

[Proof]

$$A^2 - (a+d)A = A\{A - (a+d)E\}$$

$$= \begin{pmatrix} a & b \\ c & d \end{pmatrix} \left\{ \begin{pmatrix} a & b \\ c & d \end{pmatrix} - \begin{pmatrix} a+d & 0 \\ 0 & a+d \end{pmatrix} \right\}$$

$$= \begin{pmatrix} a & b \\ c & d \end{pmatrix} \begin{pmatrix} -d & b \\ c & -a \end{pmatrix} = \begin{pmatrix} -ad+bc & 0 \\ 0 & bc-ad \end{pmatrix}$$

$$= -(ad - bc)E$$

Therefore,

$$A^2 - (a+d)A + (ad - bc)E = O.$$

Problem 7 Find A^2 and A^3 for $A = \begin{pmatrix} 2 & -3 \\ 1 & -1 \end{pmatrix}$, using the result of the above Demonstration.

5 Inverse Matrices

Given a square matrix A, if we take E as a unit matrix with the same form as A, and there exists a square matrix B which satisfies

$$AB = BA = E,$$

B is called the **inverse matrix** of A and is designated by the symbol A^{-1}.

Note: If B is the inverse matrix of A, then A is the inverse matrix of B.

Example
$$\begin{pmatrix} 2 & 1 \\ 7 & 5 \end{pmatrix} \begin{pmatrix} \frac{5}{3} & \frac{-1}{3} \\ \frac{-7}{3} & \frac{2}{3} \end{pmatrix} = \begin{pmatrix} 1 & 0 \\ 0 & 1 \end{pmatrix}$$

$$\begin{pmatrix} \frac{5}{3} & \frac{-1}{3} \\ \frac{-7}{3} & \frac{2}{3} \end{pmatrix} \begin{pmatrix} 2 & 1 \\ 7 & 5 \end{pmatrix} = \begin{pmatrix} 1 & 0 \\ 0 & 1 \end{pmatrix}$$

Therefore,

$$\begin{pmatrix} 2 & 1 \\ 7 & 5 \end{pmatrix}^{-1} = \begin{pmatrix} \frac{5}{3} & \frac{-1}{3} \\ \frac{-7}{3} & \frac{2}{3} \end{pmatrix} = \frac{1}{3} \begin{pmatrix} 5 & -1 \\ -7 & 2 \end{pmatrix}.$$

Let's consider how to find the inverse matrix of the square matrix $A = \begin{pmatrix} a & b \\ c & d \end{pmatrix}$.

Let us assume that B, the inverse matrix of A, exists, and take

$$B = \begin{pmatrix} x & u \\ y & v \end{pmatrix}.$$

Then

$$AB = \begin{pmatrix} a & b \\ c & d \end{pmatrix} \begin{pmatrix} x & u \\ y & v \end{pmatrix} = \begin{pmatrix} 1 & 0 \\ 0 & 1 \end{pmatrix}. \qquad (*)$$

Since

$$\begin{cases} ax + by = 1 & (1) \\ cx + dy = 0 & (2) \end{cases}$$

$$\begin{cases} au + bv = 0 & (3) \\ cu + dv = 1 & (4) \end{cases}$$

From (1) × d − (2) × b, $(ad - bc)x = d$ (5)

From (2) × a − (1) × c, $(ad - bc)y = -c$ (6)

From (3) × d − (4) × b, $(ad - bc)u = -b$ (7)

From (4) × a − (3) × c, $(ad - bc)v = a$ (8)

(i) For $ad - bc \neq 0$,

$$x = \frac{d}{ad - bc}, \quad u = \frac{-b}{ad - bc}$$

$$y = \frac{-c}{ad - bc}, \quad v = \frac{a}{ad - bc}.$$

Therefore,

$$B = \begin{pmatrix} x & u \\ y & v \end{pmatrix} = \frac{1}{ad - bc}\begin{pmatrix} d & -b \\ -c & a \end{pmatrix}. \qquad (**)$$

Now we have matrix B satisfying (*). We can easily verify that the matrix given in (**) satisfies not only (*), but also the following formula:

$$BA = \begin{pmatrix} x & u \\ y & v \end{pmatrix}\begin{pmatrix} a & b \\ c & d \end{pmatrix} = \begin{pmatrix} 1 & 0 \\ 0 & 1 \end{pmatrix}. \qquad (***)$$

Therefore,

$$AB = BA = E.$$

Thus, the matrix given by (**) is the inverse matrix of A.

(ii) For $ad - bc = 0$,

from (5), (6), (7), and (8), we have

$$a = b = c = d = 0.$$

This contradicts (1) and (4).

Therefore, for $ad - bc = 0$, matrix A has no inverse.

Note: The expression in (**) was obtained by using only (*). Therefore, the above conclusion shows that for square matrices A and B of dimension 2, if $AB = E$, then B is the inverse matrix of A.

Problem 1 Check by means of calculation that matrix B given by (**) satisfies (***).

Inverse Matrix

Given
$$A = \begin{pmatrix} a & b \\ c & d \end{pmatrix}$$

and taking $D = ad - bc$:

(1) if $D \neq 0$, then $A^{-1} = \dfrac{1}{D} \begin{pmatrix} d & -b \\ -c & a \end{pmatrix}$

(2) if $D = 0$, then matrix A has no inverse.

Demonstration 1 Find the inverse matrix of the following matrices:

(1) $A = \begin{pmatrix} 1 & 2 \\ 3 & 4 \end{pmatrix}$ (2) $B = \begin{pmatrix} 2 & 4 \\ 3 & 6 \end{pmatrix}$

[Solution] (1) Since $1 \cdot 4 - 2 \cdot 3 = -2 \neq 0$,

$$A^{-1} = \frac{1}{-2} \begin{pmatrix} 4 & -2 \\ -3 & 1 \end{pmatrix} = \begin{pmatrix} -2 & 1 \\ \frac{3}{2} & -\frac{1}{2} \end{pmatrix}.$$

(2) Since $2 \cdot 6 - 4 \cdot 3 = 0$, matrix B has no inverse.

Problem 2 Find the inverse matrix of the following matrices:

(1) $\begin{pmatrix} 0 & 3 \\ 2 & 1 \end{pmatrix}$ (2) $\begin{pmatrix} 2 & -1 \\ 3 & 1 \end{pmatrix}$

(3) $\begin{pmatrix} 0 & -1 \\ -1 & 0 \end{pmatrix}$ (4) $\begin{pmatrix} 1 & 1 \\ 1 & 1 \end{pmatrix}$

Demonstration 2 Prove that if square matrices A and B of dimension 2 have the inverse matrices A^{-1} and B^{-1}, then

$$(AB)^{-1} = B^{-1}A^{-1}.$$

[Proof]
$(AB)(B^{-1}A^{-1}) = A(BB^{-1})A^{-1}$
$= AEA^{-1} = AA^{-1} = E.$

Therefore, the inverse matrix of AB is $B^{-1}A^{-1}$.

Problem 3 Given $A = \begin{pmatrix} 3 & -6 \\ -2 & 5 \end{pmatrix}$ and $B = \begin{pmatrix} 2 & 5 \\ 1 & 3 \end{pmatrix}$, calculate $(AB)^{-1}$ and $B^{-1}A^{-1}$, and check that the equation in Demonstration 2 holds.

Problem 4 Give an indirect proof that if $A \neq O$, $B \neq O$, and $AB = O$, neither A nor B has an inverse matrix.

Simultaneous Linear Equations

You have already learned in junior high school how to solve simultaneous linear equations with x and y as the variables:

$$\begin{cases} ax + by = m \\ cx + dy = n \end{cases} \qquad (1)$$

Here, let's consider how to express the solutions of these equations in terms of matrices.

(1) can be written in terms of matrices as:

$$\begin{pmatrix} a & b \\ c & d \end{pmatrix} \begin{pmatrix} x \\ y \end{pmatrix} = \begin{pmatrix} m \\ n \end{pmatrix}.$$

Thus, if we take

$$A = \begin{pmatrix} a & b \\ c & d \end{pmatrix}, \quad \vec{u} = \begin{pmatrix} x \\ y \end{pmatrix}, \quad \vec{p} = \begin{pmatrix} m \\ n \end{pmatrix},$$

then

$$A\vec{u} = \vec{p}. \qquad (2)$$

If A has an inverse matrix A^{-1}, then

$$A^{-1} A \vec{u} = A^{-1} \vec{p}.$$

Thus,

$$\vec{u} = A^{-1} \vec{p}. \qquad (3)$$

Therefore, if matrix A has an inverse, the solutions of equation (2) are expressed by (3).

If matrix A has no inverse, there are cases when equation (2) has an infinite number of solutions and when it has no solutions.

3 MATRICES

Demonstration Solve the following simultaneous equations by means of equation (3).

$$\begin{cases} x - 2y = -1 \\ 2x + 3y = 12 \end{cases}$$

[Solution] If we take

$$A = \begin{pmatrix} 1 & -2 \\ 2 & 3 \end{pmatrix}$$

then

$$1 \cdot 3 - (-2) \cdot 2 = 7 \neq 0.$$

Therefore, from the formula for an inverse matrix on page 94,

$$A^{-1} = \frac{1}{7} \begin{pmatrix} 3 & 2 \\ -2 & 1 \end{pmatrix}.$$

Thus,

$$\begin{pmatrix} x \\ y \end{pmatrix} = \frac{1}{7} \begin{pmatrix} 3 & 2 \\ -2 & 1 \end{pmatrix} \begin{pmatrix} -1 \\ 12 \end{pmatrix} = \frac{1}{7} \begin{pmatrix} 21 \\ 14 \end{pmatrix} = \begin{pmatrix} 3 \\ 2 \end{pmatrix}.$$

Answer: $x = 3, y = 2$

Problem 1 Find the inverse of the matrix $\begin{pmatrix} 3 & 2 \\ 1 & 4 \end{pmatrix}$, and then use it to solve the following simultaneous equations.

(1) $\begin{cases} 3x + 2y = 7 \\ x + 4y = 9 \end{cases}$ (2) $\begin{cases} 3x + 2y = 51 \\ x + 4y = 47 \end{cases}$

(3) $\begin{cases} 3x + 2y = 89 \\ x + 4y = -47 \end{cases}$ (4) $\begin{cases} 3x + 2y = 311 \\ x + 4y = 707 \end{cases}$

Problem 2 Given $A = \begin{pmatrix} 1 & -2 \\ -3 & 6 \end{pmatrix}$ and $\vec{p} = \begin{pmatrix} -3 \\ 9 \end{pmatrix}$, show that $\vec{u} = \begin{pmatrix} -1 + 2t \\ 1 + t \end{pmatrix}$ is the solution of $A\vec{u} = \vec{p}$ for any arbitrary real number t.

Problem 3 Given $A = \begin{pmatrix} 1 & -2 \\ -3 & 6 \end{pmatrix}$ and $\vec{p} = \begin{pmatrix} -1 \\ 4 \end{pmatrix}$, show that the equation $A\vec{u} = \vec{p}$ has no solutions.

Exercises

1. Define the values of $a, b, c,$ and d such that the following equality holds:
$$\begin{pmatrix} a & 6 \\ 1 & b \end{pmatrix} + \frac{1}{2}\begin{pmatrix} 2 & d-1 \\ 3 & 4 \end{pmatrix} = \begin{pmatrix} 5 & 7 \\ 2c & -3 \end{pmatrix}.$$

2. Given $A = \begin{pmatrix} 2 & 4 \\ -3 & 7 \end{pmatrix}, B = \begin{pmatrix} 1 & -1 \\ 2 & -3 \end{pmatrix}$, and $C = \begin{pmatrix} 1 & 0 \\ 6 & -5 \end{pmatrix}$.

    (1) Calculate $2(A - 3B) - 3(A - C)$.

    (2) Find matrix X which satisfies $A - X = 3(A - B) + 2(X + 2C)$.

3. We need to choose two different matrices from among $A = \begin{pmatrix} 1 & -2 \end{pmatrix}$, $B = \begin{pmatrix} 3 \\ 2 \end{pmatrix}$, and $C = \begin{pmatrix} -2 & 1 \\ 4 & 0 \end{pmatrix}$ and find their product. When the product can be found, calculate it.

4. Given $A = \begin{pmatrix} 1 & 2 \\ 0 & -1 \end{pmatrix}$ and $B = \begin{pmatrix} 1 & -1 \\ -2 & 1 \end{pmatrix}$, calculate the following matrices:

    (1) A^2 (2) $A^2 - B^2$

    (3) $(A + B)(A - B)$ (4) $A^2 + 2AB + B^2$

5. Check that the following equation holds:
$$\begin{pmatrix} 0 & 1 \\ -1 & 0 \end{pmatrix}^2 + \begin{pmatrix} -1 & 0 \\ 0 & -1 \end{pmatrix}^2 = \begin{pmatrix} 0 & 0 \\ 0 & 0 \end{pmatrix}.$$

6. Calculate $\begin{pmatrix} x & y \end{pmatrix}\begin{pmatrix} a & h \\ h & b \end{pmatrix}\begin{pmatrix} x \\ y \end{pmatrix}$.

3 MATRICES

7. Given a 2 × 2 matrix $A = \begin{pmatrix} a & b \\ c & d \end{pmatrix}$, we will take A' as the matrix $\begin{pmatrix} a & c \\ b & d \end{pmatrix}$ created by interchanging the rows and columns. If A and B are 2 × 2 matrices, prove the following equations.

   (1) $(A + B)' = A' + B'$, $(AB)' = B'A'$

   (2) For $ad - bc \neq 0$, $(A^{-1})' = (A')^{-1}$

8. Given a matrix $A = \begin{pmatrix} a & b \\ c & d \end{pmatrix}$, prove that if $a + d = 1$ and $ad - bc = 0$, then $A^2 = A$.

9. Find matrix X, given $A = \begin{pmatrix} 2 & 4 \\ 3 & 7 \end{pmatrix}$ and $B = \begin{pmatrix} 4 & 10 \\ 5 & 3 \end{pmatrix}$:

   (1) $AX = B$ \hspace{2cm} (2) $XA = B$

10. Define the value of x such that the matrix $\begin{pmatrix} 2-x & 4 \\ 5 & 3-x \end{pmatrix}$ has no inverse.

2 LINEAR TRANSFORMATIONS

1 The Meaning of a Linear Transformation

Movements of points, such as reflecting each point on a coordinate plane with respect to the origin or the coordinate axes, or rotating it about the origin, can be expressed using matrices.

Example 1 A movement such that each point $P(x, y)$ on the xy-plane moves to the point $Q(x', y')$, symmetric with respect to the x-axis, is referred to as a "reflection with respect to the x-axis."

The relation between the coordinates of points P and Q is

$$x' = x, \quad y' = -y.$$

That is,

$$\begin{cases} x' = 1 \cdot x + 0 \cdot y \\ y' = 0 \cdot x + (-1) \cdot y \end{cases}$$

This relation can be expressed as a matrix:

$$\begin{pmatrix} x' \\ y' \end{pmatrix} = \begin{pmatrix} 1 & 0 \\ 0 & -1 \end{pmatrix} \begin{pmatrix} x \\ y \end{pmatrix}.$$

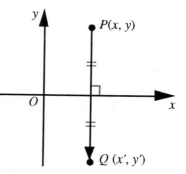

Problem 1 If the point (x, y) moves to the point (x', y') by the following reflections, express the relations between x, y and x', y' using a matrix as in Example 1:

(1) reflection with respect to the y-axis

(2) reflection with respect to the origin

(3) reflection with respect to the straight line $y = x$.

Example 2 If k is a non-zero constant, movement of each point $P(x, y)$ on a plane to point $Q(kx, ky)$ is referred to as a **similar transformation** with a similarity ratio of k and the origin as the center.

This transformation causes point $P(x, y)$ to move to point $Q(x', y')$, and then

$$\begin{cases} x' = kx \\ y' = ky \end{cases}.$$

Thus,

$$\begin{pmatrix} x' \\ y' \end{pmatrix} = \begin{pmatrix} k & 0 \\ 0 & k \end{pmatrix} \begin{pmatrix} x \\ y \end{pmatrix}.$$

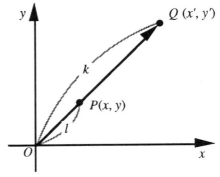

In Example 2, if $k = 1$, then

$$x' = x, \quad y' = y.$$

Therefore, each point on the plane moves to itself. That means each point on the plane does not move. This is called an **identity transformation**.

Example 3 Movement of each point $P(x, y)$ on a plane to point $P'(x', y')$ given by rotating point P by angle θ about the origin O is referred to as **rotation** of angle θ about the origin.

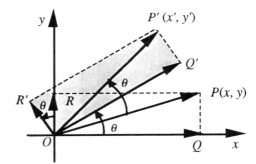

Draw lines *PQ* and *PR* from point *P* perpendicular to the *x*- and *y*-axes. Rectangle *OQPR* becomes the congruent rectangle *OQ'P'R'* by rotation of angle θ, and then

$\overrightarrow{OQ} = (x, 0)$

$\overrightarrow{OR} = (0, y)$

$OQ' = (x \cos \theta, x \sin \theta)$

$OR' = (-y \sin \theta, y \cos \theta)$.

And since

$\overrightarrow{OP'} = \overrightarrow{OQ'} + \overrightarrow{OR'}$

$(x', y') = (x \cos \theta, x \sin \theta) + (-y \sin \theta, y \cos \theta)$

Therefore,

$$\begin{cases} x' = x \cos \theta - y \sin \theta \\ y' = x \sin \theta + y \cos \theta \end{cases}$$

Thus,

$$\begin{pmatrix} x' \\ y' \end{pmatrix} = \begin{pmatrix} \cos \theta & -\sin \theta \\ \sin \theta & \cos \theta \end{pmatrix} \begin{pmatrix} x \\ y \end{pmatrix}.$$

Problem 2 Find the coordinates of the points to which the point (2, 1) moves by rotation around the origin of the following angles:

(1) 30°   (2) 45°   (3) 120°

The movements in Examples 1, 2, and 3 all involve movement of the point (x, y) on a plane to the point (x', y'), and all of them can be given as a linear expression in x and y. That is, taking $a, b, c,$ and d as constants,

$$\begin{cases} x' = ax + by \\ y' = cx + dy \end{cases} \quad (1)$$

Movement expressed by a linear expression without a constant term is called a **linear transformation**.

Linear transformations are designated by letters such as f or g.

Take f as the linear transformation expressed by the equations in (1) above. If we express (1) as a matrix, then

$$\begin{pmatrix} x' \\ y' \end{pmatrix} = \begin{pmatrix} a & b \\ c & d \end{pmatrix} \begin{pmatrix} x \\ y \end{pmatrix}.$$

Thus, if we take

$$A = \begin{pmatrix} a & b \\ c & d \end{pmatrix}, \ \vec{p} = \begin{pmatrix} x \\ y \end{pmatrix}, \ \vec{p'} = \begin{pmatrix} x' \\ y' \end{pmatrix},$$

then,

$$\vec{p'} = A\vec{p}.$$

This matrix A is called the **matrix of linear transformation** f.

When the point (x, y) moves to the point (x', y') under a linear transformation f, the point (x', y') is called the **image** of the point (x, y) under f.

The image of the origin under a linear transformation is the origin.

Problem 3 Given that the matrix of linear transformation f is $\begin{pmatrix} 2 & 1 \\ 3 & 4 \end{pmatrix}$, find the image of the following points under f.

$(1, 0), \ (0, 1), \ (-2, 4), \ (5, -6)$

  The Linearity of Linear Transformations

A linear transformation f which causes a point (x, y) to move to the point (x', y') can be regarded as

the correspondence of vector $\vec{p} = \begin{pmatrix} x \\ y \end{pmatrix}$ to vector $\vec{p'} = \begin{pmatrix} x' \\ y' \end{pmatrix}$.

This correspondence can be expressed as

$$\vec{p'} = f(\vec{p})$$

and vector $\vec{p'}$ is called the **image** of vector $\vec{p}$ under f.

Problem 1 Given that the matrix of linear transformation f is $\begin{pmatrix} a & b \\ c & d \end{pmatrix}$, show that the images of the basic vectors $\vec{e_1} = \begin{pmatrix} 1 \\ 0 \end{pmatrix}$ and $\vec{e_2} = \begin{pmatrix} 0 \\ 1 \end{pmatrix}$ are $\vec{e_1}' = \begin{pmatrix} a \\ c \end{pmatrix}$ and $\vec{e_2}' = \begin{pmatrix} b \\ d \end{pmatrix}$, respectively.

Problem 2 If the images of the basic vectors $\vec{e_1}$ and $\vec{e_2}$ under linear transformation f are $\begin{pmatrix} 2 \\ 1 \end{pmatrix}$ and $\begin{pmatrix} -1 \\ 3 \end{pmatrix}$, find the matrix of f.

When we consider linear transformation f as the correspondence of one vector to another vector, the following properties hold. These properties represent the **linearity** of a linear transformation.

The Linearity of a Linear Transformation

(I) $f(\vec{p_1} + \vec{p_2}) = f(\vec{p_1}) + f(\vec{p_2})$

(II) $f(k\vec{p}) = kf(\vec{p})$

[Proof] Take A as the matrix of linear transformation f.

$$f(\vec{p_1} + \vec{p_2}) = A(\vec{p_1} + \vec{p_2})$$
$$= A\vec{p_1} + A\vec{p_2} = f(\vec{p_1}) + f(\vec{p_2})$$
$$f(k\vec{p}) = A(k\vec{p}) = k(A\vec{p}) = kf(\vec{p})$$

Problem 3 Show that the following equality holds for linear transformation f:

$$f(k\vec{p_1} + l\vec{p_2}) = kf(\vec{p_1}) + lf(\vec{p_2})$$

Take $\vec{e_1}'$ and $\vec{e_2}'$ as the images of the basic vectors $\vec{e_1}$ and $\vec{e_2}$ under linear transformation f, and then the image $\vec{p}' = f(\vec{p})$ under f of an arbitrary vector in a plane

$$\vec{p} = \begin{pmatrix} x \\ y \end{pmatrix} = x\begin{pmatrix} 1 \\ 0 \end{pmatrix} + y\begin{pmatrix} 0 \\ 1 \end{pmatrix} = x\vec{e_1} + y\vec{e_2}$$

by the linearity of a linear transformation is

$$\vec{p}' = x\vec{e_1}' + y\vec{e_2}'.$$

The figure below shows the case for $\vec{p} = 3\vec{e_1} + 2\vec{e_2}$.

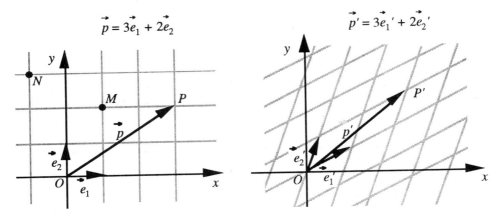

Problem 4 To which points in the figure to the right do the two points M and N in the left-hand figure move under this linear transformation?

Demonstration Under linear transformation f, two different points A and B move to two different points A' and B'. Prove that the image M' of the midpoint M of line segment AB is the midpoint of line segment $A'B'$.

[**Proof**] Take $\vec{a}, \vec{b}$ and $\vec{m}$ as the position vectors of A, B, and M, respectively. Then the position vectors of A', B', and M' are $f(\vec{a}), f(\vec{b})$, and $f(\vec{m})$. Since M is the midpoint of AB,

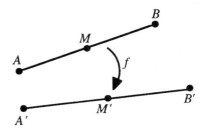

$$\vec{m} = \frac{\vec{a} + \vec{b}}{2}.$$

From the linearity of f,

$$f(\vec{m}) = f(\frac{\vec{a} + \vec{b}}{2})$$

$$= \frac{f(\vec{a} + \vec{b})}{2} = \frac{f(\vec{a}) + f(\vec{b})}{2}.$$

Therefore, M' is the midpoint of line segment $A'B'$.

Problem 5 Under the same assumptions as in the above Demonstration, prove that the image P' of point P, which divides line segment AB internally at a ratio of $m : n$, divides line segment $A'B'$ internally at a ratio of $m : n$.

  Linear Transformations and Figures

Take A as the matrix of a linear transformation f, and take the point (x', y') as the image of an arbitrary point (x, y) on a plane under f,

$$\begin{pmatrix} x' \\ y' \end{pmatrix} = A \begin{pmatrix} x \\ y \end{pmatrix}, \qquad (1)$$

If A has an inverse matrix A^{-1}, we can multiply both sides of (1) by A^{-1} from left to right, and reverse sides, and then we obtain

$$\begin{pmatrix} x \\ y \end{pmatrix} = A^{-1} \begin{pmatrix} x' \\ y' \end{pmatrix}. \qquad (2)$$

Equalities (2) and (1) are equivalent. Therefore, each point (x', y') on a plane is the image of single point (x, y) given by (2).

This fact allows us to formulate the following generalization.

> If the matrix of a linear transformation f has an inverse matrix, under f the entire plane moves on the entire plane, and different points move to different points.

Problem 1 Find the point which moves to the point (7, 5) under the linear transformation whose matrix is $\begin{pmatrix} 1 & 2 \\ 3 & 4 \end{pmatrix}$.

If the matrix of a linear transformation f is not a zero matrix and has no inverse, under f the entire plane moves to one straight line through the origin.

Example If the point (x, y) moves to the point (x', y') under the linear transformation f whose matrix is $\begin{pmatrix} 2 & 1 \\ 4 & 2 \end{pmatrix}$,

$$\begin{cases} x' = 2x + y \\ y' = 4x + 2y \end{cases}$$

Therefore,

$$y' = 2x.$$

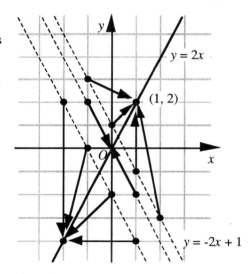

Thus, the entire plane moves to the straight line $y = 2x$ under f.

3 MATRICES

Problem 2 Show that the set of points whose image under the linear transformation in the above Example is the point (1, 2) on the line $y = 2x$ is the straight line $2x + y = 1$.

The Image of a Straight Line

If two different points A and B move to two different points A' and B' under linear transformation f, then straight line AB moves to straight line $A'B'$.

[Proof] Take $\vec{a}, \vec{b}, \vec{a'}$, and $\vec{b'}$ as the position vectors of points A, B, A', and B', and then

$$\vec{a'} = f(\vec{a}), \quad \vec{b'} = f(\vec{b}).$$

Take $\vec{p}$ as the position vector of point P moving along line AB, and then using parameter t, straight line AB can be expressed as

$$\vec{p} = \vec{a} + t(\vec{b} - \vec{a}).$$

Take $\vec{p'}$ as the position vector of P', the image of P, and then

$$\vec{p'} = f(\vec{p}).$$

From the linearity of f,

$$\vec{p'} = \vec{a'} + t(\vec{b'} - \vec{a'}).$$

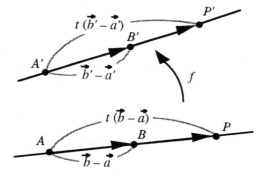

Therefore, the set of points P' whose position vector is $\vec{p'}$ is straight line $A'B'$.

Problem 3 If two different points A and B move to two different points A' and B' under linear transformation f, to what figure does line segment AB move?

If two different points A and B move to the same point A', all the points on straight line AB move to a single point A'.

Problem 4 Prove the above statement.

Demonstration 1

Take l as the straight line $x + 3y - 3 = 0$. To what figure does l move under the linear transformation expressed under the matrix $\begin{pmatrix} 1 & -1 \\ 2 & 4 \end{pmatrix}$?

[Solution] Let's consider the images of two different points $(0, 1)$ and $(3, 0)$ on line l.

$$\begin{pmatrix} 1 & -1 \\ 2 & 4 \end{pmatrix}\begin{pmatrix} 0 \\ 1 \end{pmatrix} = \begin{pmatrix} -1 \\ 4 \end{pmatrix}$$

$$\begin{pmatrix} 1 & -1 \\ 2 & 4 \end{pmatrix}\begin{pmatrix} 3 \\ 0 \end{pmatrix} = \begin{pmatrix} 3 \\ 6 \end{pmatrix}$$

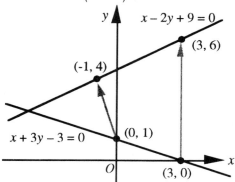

Therefore, line l moves to the straight line through the two points $(-1, 4)$ and $(3, 6)$

$$x - 2y + 9 = 0.$$

[Alternate Solution] Assume that the point (x, y) moves to the point (x', y'), and then

$$\begin{pmatrix} x' \\ y' \end{pmatrix} = \begin{pmatrix} 1 & -1 \\ 2 & 4 \end{pmatrix}\begin{pmatrix} x \\ y \end{pmatrix}. \tag{1}$$

Multiplying both sides of (1) by $\begin{pmatrix} 1 & -1 \\ 2 & 4 \end{pmatrix}^{-1} = \frac{1}{6}\begin{pmatrix} 4 & 1 \\ -2 & 1 \end{pmatrix}$ from left to right and expressing x and y in terms of x' and y', we obtain

$$x = \frac{4x' + y'}{6}, \quad y = \frac{2x' + y'}{6}. \tag{2}$$

The condition that the point (x, y) lies on the straight line $x + 3y - 3 = 0$ is formed by substituting (2) in the equation of the line,

$$\frac{4x' + y'}{6} + 3 \cdot \frac{2x' + y'}{6} - 3 = 0.$$

Rearranging,

$$x' - 2y' + 9 = 0.$$

Therefore, the line $x + 3y - 3 = 0$ moves to the line

$$x - 2y + 9 = 0.$$

Problem 5 To what figures do the following straight lines move under the linear transformation expressed by the matrix $\begin{pmatrix} 1 & 2 \\ 3 & 4 \end{pmatrix}$?

(1) $y = 2x + 1$ (2) $2x + y + 1 = 0$ (3) the x-axis

Problem 6 To what figure do the following straight lines move under the linear transformation expressed by the matrix $\begin{pmatrix} 4 & -2 \\ -6 & 3 \end{pmatrix}$?

(1) $y = 3x - 2$ (2) the y-axis (3) $2x - y + 1 = 0$

Demonstration 2 If the matrix of linear transformation f is $\begin{pmatrix} 3 & -5 \\ 2 & -4 \end{pmatrix}$:

(1) find a point that does not move under f.

(2) find a straight line that moves to itself under f.

[Solution] (1) Take (x, y) as the point we want to find, and then

$$x = 3x - 5y, \quad y = 2x - 4y.$$

Hence,

$$y = \frac{2}{5}x.$$

Therefore, the point we want to find is represented by all the points on the straight line $y = \frac{2}{5}x$.

(2) Tal e the equation of the straight line l we want to find as

$$y = mx + n.$$

If we find the image of two of the points $(0, n)$ and $(1, m + n)$ on l under f, we obtain, respectively,

$$(-5n, -4n), \quad (3 - 5m - 5n, 2 - 4m - 4n). \tag{1}$$

Since l moves to itself under f, both points in (1) lie on l. That is,

$$\begin{cases} -4n = m(-5n) + n & (2) \\ 2 - 4m - 4n = m(3 - 5m - 5n) + n. & (3) \end{cases}$$

From (3) − (2),

$$2 - 4m = m(3 - 5m). \qquad (4)$$

Solving (4) with respect to m, we obtain

$$m = 1 \text{ or } m = \frac{2}{5}.$$

For $m = 1$, from (2), n is an arbitrary constant.
For $m = \frac{2}{5}$, from (2), $n = 0$.

Therefore, the line we want to find, for an arbitrary constant n, is

$$\text{either } y = x + n \text{ or } y = \frac{2}{5}x.$$

Problem 7 To what straight line does the line $x = c$ move under linear transformation f in Demonstration 2?

Problem 8 Find the points that do not move under the linear transformations expressed by the following matrices. Are any straight lines transformed to themselves? If so, find them.

(1) $\begin{pmatrix} 2 & 1 \\ 2 & 3 \end{pmatrix}$ (2) $\begin{pmatrix} 5 & -6 \\ 3 & -4 \end{pmatrix}$ (3) $\begin{pmatrix} 1 & -1 \\ 2 & 3 \end{pmatrix}$

Demonstration 3 Find the equation of the curve created by rotating the hyperbola $x^2 - y^2 = 2$ about the origin by 45°.

[Solution] Take $P'(x', y')$ as the point created by rotating the point $P(x, y)$ by 45° about the origin; then the formula on page 102 gives us

$$\begin{cases} x' = x \cos 45° - y \sin 45° = \frac{1}{\sqrt{2}}(x - y) \\ y' = x \sin 45° + y \cos 45° = \frac{1}{\sqrt{2}}(x + y) \end{cases}$$

Solving with respect to x and y,

$$\begin{cases} x = \frac{1}{2}(\sqrt{2}x' + \sqrt{2}y') = \frac{1}{\sqrt{2}}(x' + y') \\ y = \frac{1}{2}(-\sqrt{2}x' + \sqrt{2}y') = \frac{1}{\sqrt{2}}(-x' + y') \end{cases}$$

If we express the condition that the point $P(x, y)$ lies on the hyperbola $x^2 - y^2 = 2$ in terms of x' and y', we get

$$\frac{1}{2}(x' + y')^2 - \frac{1}{2}(-x' + y')^2 = 2.$$

That is, $\qquad\qquad\qquad\qquad x'y' = 1.$

Therefore, the curve we want to find is $xy = 1$.

Problem 9 Find the equation of the straight line created by rotating the line $x + y = 1$ by $60°$ about the origin.

The Composite and Inverse of Linear Transformations

The Composite of Linear Transformations

Take A and B as the matrices of two linear transformations f and g. Take $\vec{p}'$ as the image of vector $\vec{p}$ under f and $\vec{p}''$ as the image of $\vec{p}'$ under g. Then

$$\vec{p}' = A\vec{p}$$
$$\vec{p}'' = B\vec{p}'.$$

Therefore,

$$\vec{p}'' = B(A\vec{p}) = (BA)\vec{p}.$$

Since BA is a matrix, the correspondence from $\vec{p}$ to $\vec{p}''$ of the successive linear transformations f and g is also a linear transformation. This linear transformation is called the composite transformation of f and g, and is designated by the symbol $g \circ f$.

The Composite of a Linear Transformation

If A and B are the matrices of linear transformations f and g, then the matrix of the composite transformation $g \circ f$ is the product BA of B and A.

Problem 1 If the matrices of linear transformations f and g are $\begin{pmatrix} 1 & 2 \\ 0 & -1 \end{pmatrix}$ and $\begin{pmatrix} 3 & 0 \\ 1 & -1 \end{pmatrix}$, find the matrices of the composite transformations $g \circ f$ and $f \circ g$.

Problem 2 If $\begin{pmatrix} x' \\ y' \end{pmatrix} = \begin{pmatrix} 1 & -2 \\ 3 & 4 \end{pmatrix}\begin{pmatrix} x \\ y \end{pmatrix}$, $\begin{pmatrix} x'' \\ y'' \end{pmatrix} = \begin{pmatrix} 3 & 1 \\ 4 & -1 \end{pmatrix}\begin{pmatrix} x' \\ y' \end{pmatrix}$

express x'' and y'' in terms of x and y.

Problem 3 Show that the composite of a reflection with respect to the x-axis and a reflection with respect to the line $y = x$ is a rotation of 90° about the origin.

[Reference:] The Addition Theorems for Trigonometric Functions

The addition theorems for trigonometric functions are studied in *Basic Analysis*. But here let's derive these theorems using matrices.

The composite transformation of a rotation of angle α about the origin and a rotation of angle β about the origin is a rotation of angle $\alpha + \beta$. Therefore, the following relation holds between the matrices that express these rotations:

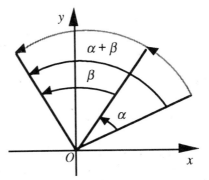

$$\begin{pmatrix} \cos(\alpha+\beta) & -\sin(\alpha+\beta) \\ \sin(\alpha+\beta) & \cos(\alpha+\beta) \end{pmatrix} = \begin{pmatrix} \cos\beta & -\sin\beta \\ \sin\beta & \cos\beta \end{pmatrix}\begin{pmatrix} \cos\alpha & -\sin\alpha \\ \sin\alpha & \cos\alpha \end{pmatrix}$$

Comparing the first row of both sides, we obtain the following formulas for $\cos(\alpha+\beta)$ and $\sin(\alpha+\beta)$.

$$\cos(\alpha+\beta) = \cos\alpha\cos\beta - \sin\alpha\sin\beta$$

$$\sin(\alpha+\beta) = \sin\alpha\cos\beta + \cos\alpha\sin\beta$$

These are the **addition theorems** for sine and cosine.

The Inverse of a Linear Transformation

Take A as the matrix of a linear transformation f, and take $\vec{p}'$ as the image of $\vec{p}$ under f. Then

$$\vec{p}' = A\vec{p}.$$

If matrix A has an inverse A^{-1}, then by multiplying both sides of this equality by A^{-1} from left to right, we obtain

$$\vec{p} = A^{-1}\vec{p}'.$$

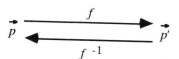

Therefore, the reverse correspondence of f from $\vec{p}'$ to $\vec{p}$ is also a linear transformation, which is called the **inverse transformation** of f and is designated by f^{-1}.

The Inverse of a Linear Transformation

If matrix A of a linear transformation has an inverse A^{-1}, the inverse transformation of f is also a linear transformation, and its matrix is A^{-1}.

Problem 4 Find the matrix that expresses the inverse of the linear transformation expressed by each matrix:

(1) $\begin{pmatrix} x' \\ y' \end{pmatrix} = \begin{pmatrix} 1 & -2 \\ 0 & 1 \end{pmatrix} \begin{pmatrix} x \\ y \end{pmatrix}$

(2) $\begin{pmatrix} x' \\ y' \end{pmatrix} = \begin{pmatrix} 3 & 9 \\ 1 & 2 \end{pmatrix} \begin{pmatrix} x \\ y \end{pmatrix}$

Problem 5 Take f^{-1} as the inverse of a linear transformation f, and then check that the composite transformations $f^{-1} \circ f$ and $f \circ f^{-1}$ are both identity transformations.

Mapping

The linear transformations we have learned about so far, as well as the functions we studied in the grade 10 textbook, are special cases of what is called "mapping". Now let's learn about mapping.

The Meaning of Mapping

Take X and Y as two sets. If every element x of X corresponds to a single element y of Y, this correspondence is called a **mapping** from X to Y.

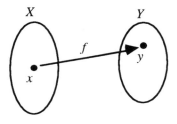

A mapping is generally designated by a letter such as f or g, and we designate that f is a mapping from X to Y by

$$f \colon X \to Y.$$

If element x of X corresponds to element y of Y under a mapping f from X to Y, y is called the **image** of x under f, and is written as

$$y = f(x).$$

This can also be written as $f \colon x \to y$ or simply as $x \to y$.

Example 1 Take R as the set of all real numbers. Then the quadratic function

$$y = x^2 + 1$$

defines one mapping from R to R. If we take this mapping as f, and then for any arbitrary real number x, it is

$$f(x) = x^2 + 1.$$

Example 2 Take S as the set of all the points on the xy-plane. If A is a 2×2 matrix, the linear transformation represented under A

$$\begin{pmatrix} x' \\ y' \end{pmatrix} = A \begin{pmatrix} x \\ y \end{pmatrix} \tag{1}$$

makes each element (x, y) of S correspond to an element (x', y') of S. Therefore, this transformation is a mapping from S to S.

Take V as the set of all vectors on a plane. (1) also makes the element $\vec{p} = \begin{pmatrix} x \\ y \end{pmatrix}$ of V correspond to $\vec{p'} = \begin{pmatrix} x' \\ y' \end{pmatrix}$, and therefore it is also considered a mapping from V to V.

Example 3 Given $X = \{x | x \text{ is a positive integer}\}$ and $Y = \{0, 1\}$. A mapping establishes a relation between x, an arbitrary element of X, and

the remainder when x is divided by 2

is a mapping from X to Y.

Problem 1 Taking f as the mapping in Example 3, find a positive integer x such that $f(x) = 0$. Then find a positive integer x such that $f(x) = 1$.

Problem 2 Take X as the set of all positive integers and take f as a mapping from X to X that makes each positive integer x correspond to the number of divisors it has. Find $f(6)$ and $f(12)$.

The Domain and Range of a Mapping

Given a mapping from X to Y

$f: X \to Y$

X is called the **domain** of f.

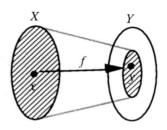

The set $\{f(x) | x \in X\}$ is called the **range** of f. Thus, the range of f is the set of all images of the elements of X.

The domain and the range of a function, which you learned about in Mathematics I (grade 10), are identical to the domain and range when a function is considered as a mapping.

116 3 MATRICES

Problem 3 Take S as the set of all the points in the xy-plane. Find the range of the linear transformation $f: S \to S$ expressed by the following matrices:

(1) $\begin{pmatrix} 1 & 2 \\ 3 & 4 \end{pmatrix}$ (2) $\begin{pmatrix} 1 & 2 \\ -3 & -6 \end{pmatrix}$

The Composite of Mappings

Given sets X, Y, and Z, and the mappings

$f: X \to Y$

$g: Y \to Z$

the mapping from X to Z that makes each element of X correspond to

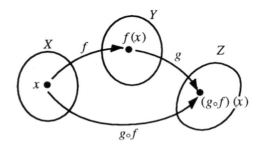

$g(f(x))$ = the image under g of the image of x under f

is called the **composite mapping**, and is designated by $g \circ f$. That is,

$$(g \circ f)(x) = g(f(x)).$$

The composite transformation we have already studied is a special case of a composite mapping.

Example 4 Take R as the set of all real numbers, and $f: R \to R$ and $g: R \to R$ are defined by

$$f(x) = 2x + 1, \qquad g(x) = x^2.$$

In this case

$$\begin{aligned}(g \circ f)(x) &= g(f(x)) \\ &= g(2x+1) = (2x+1)^2.\end{aligned}$$

Problem 4 Find an expression which gives the composite mapping for f and g in Example 4.

As you can see from the result of Example 4 and Problem 4, even if $g \circ f$ and $f \circ g$ are both defined, they are not generally the same.

However, given four sets A, B, C, and D and three mappings

$$f_1: A \to B, \quad f_2: B \to C, \quad f_3: C \to D,$$

the mappings from A to D

$$f_3 \circ (f_2 \circ f_1), \quad (f_3 \circ f_2) \circ f_1$$

are the same mapping. That is,

the associative law holds for composite mapping.

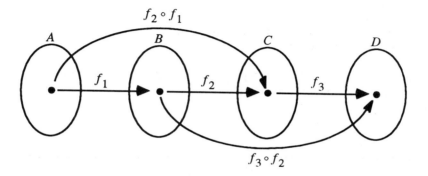

Problem 5 Take R as the set of all real numbers. Given three mappings from R to R by $f_1(x) = x + 1$, $f_2(x) = x^2$, and $f_3(x) = x - 3$, check that both $f_3 \circ (f_2 \circ f_1)$ and $(f_3 \circ f_2) \circ f_1$ are equivalent to the mapping f defined by

$$f(x) = (x + 1)^2 - 3.$$

Inverse Mapping

The mapping $f: X \to Y$ has the following two properties:

(1) The range of f is identical to Y.

(2) Different elements of X correspond to different elements of Y. That is,

$$x_1 \neq x_2 \to f(x_1) \neq f(x_2).$$

In this situation, for each element y of Y there is only one element x of X such that

$$f(x) = y$$

is defined.

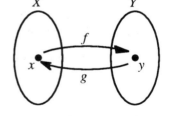

Mapping g, which makes y correspond to x, is a mapping from Y to X such that

$$y = f(x) \leftrightarrow x = g(y).$$

This relation g is called the **inverse mapping** of f and is designated by f^{-1}.

The inverse of a function, which you learned in Mathematics I, and the inverse of a linear transformation, which you have learned about in this chapter, are both special cases of inverse mapping.

A mapping from X to X that makes each element of set X correspond to x itself is called an **identity mapping** of X.

If $f: X \to Y$ has the inverse mapping $f^{-1}: Y \to X$, $f^{-1} \circ f$ is the identity mapping of X, and $f \circ f^{-1}$ is the identity mapping of Y.

Exercises

1. Is each transformation that moves the point $P(x, y)$ on a plane to the point $P'(x', y')$ specified below a linear transformation or not? If it is a linear transformation, find its matrix.

   (1) Point P' is the point symmetric to point P with respect to the line $x = 1$.

   (2) Point P' is the point at which a perpendicular line from point P to the x-axis intersects the x-axis.

   (3) Point P' is the point given by translating the point P by the vector $(1, -2)$.

   (4) Point P' is the point at which a line through the point P with a slope of 2 intersects the line $y = x$.

2. Take f as a linear transformation that moves the two points (1, 0) and (0, 1) to the two points (-1, 2) and (3, 1).

   (1) Find the matrix of f.

   (2) Find the image of the point (-3, 2) under f.

   (3) Find the image of the point (5, -3) under the inverse transformation of f.

3. Given linear transformation f, find $f(3\vec{u} - 2\vec{v})$ for $f(\vec{u}) = \begin{pmatrix} 1 \\ 2 \end{pmatrix}$ and $f(\vec{v}) = \begin{pmatrix} -2 \\ 1 \end{pmatrix}$.

4. To what figure does the entire plane move under the linear transformations expressed by the following matrices:

   (1) $\begin{pmatrix} 2 & -3 \\ 1 & 1 \end{pmatrix}$ (2) $\begin{pmatrix} 2 & 1 \\ 6 & 3 \end{pmatrix}$ (3) $\begin{pmatrix} 0 & 0 \\ 0 & 0 \end{pmatrix}$

5. Find the figure to which the straight line $2x - y + 3 = 0$ moves under the linear transformation $(x, y) \to (3x - y, -2x + y)$.

6. Find the matrix of a linear transformation that moves the two points (2, 1) and (-1, 3) to (8, -5) and (-11, 6), respectively.

7. Find the image of the point (-6, 7) under composite transformation $g \circ f^{-1}$, given that the matrices of linear transformations f and g are $\begin{pmatrix} 1 & 2 \\ -1 & -3 \end{pmatrix}$ and $\begin{pmatrix} -2 & 3 \\ 1 & -2 \end{pmatrix}$, respectively.

Chapter Exercises

A

1. Find values of x and y such that $x\begin{pmatrix} 1 & 0 \\ 0 & 1 \end{pmatrix} + y\begin{pmatrix} 2 & 1 \\ 0 & 1 \end{pmatrix} = \begin{pmatrix} 2 & 1 \\ 0 & 1 \end{pmatrix}^3$ holds.

2. Find values of x and y such that the matrix $\begin{pmatrix} x & 5 \\ -7 & y \end{pmatrix}$ is also the inverse of itself.

3. Determine the values of the constant k such that the simultaneous equations
$$\begin{cases} 2x + 3y = kx \\ 4x + 3y = ky \end{cases}$$
have solutions other than $x = y = 0$.

4. Show that
$$ps - qr = (ad - bc)(a'd' - b'c')$$
if $\begin{pmatrix} a & b \\ c & d \end{pmatrix}\begin{pmatrix} d & b' \\ c' & d' \end{pmatrix} = \begin{pmatrix} p & q \\ r & s \end{pmatrix}$.

5. Find the matrix of a similar transformation that moves the parabola $y = ax^2$ to the parabola $y = x^2$.

6. Find the matrix of a linear transformation that moves all the points on the line $3x + 2y - 1 = 0$ to the point $(-1, 1)$.

7. Find the matrix of a linear transformation that moves all the points on the plane to the straight line $y = 2x$.

8. Take l as a straight line through the origin such that the angle it forms with the positive ray of the x-axis is θ.

    (1) Take P' as the point given by reflecting the point P with respect to the x-axis and rotating it by an angle of 2θ about the origin. Show that P' is located in a position symmetric to P with respect to the line l.

(2) Find the matrix of a linear transformation that expresses a reflection with respect to the line l on a plane.

B

1. Show that matrix A, such that $XA = AX$ for any square matrix of dimension 2, has the form $A = kE$, where k is any number and E is a unit matrix of dimension 2.

2. The parametric representation of line l is $\begin{pmatrix} x \\ y \end{pmatrix} = \begin{pmatrix} 4 \\ 3 \end{pmatrix} + t \begin{pmatrix} 2 \\ 1 \end{pmatrix}$. To what figure does l move under the linear transformation expressed by the matrix $\begin{pmatrix} 4 & -3 \\ -2 & 2 \end{pmatrix}$?

3. Take $\begin{pmatrix} -2 & a \\ a & 0 \end{pmatrix}$ as the matrix of linear transformation f, where $a \neq 0$.

   (1) To what figures do the x- and y-axes move under f?

   (2) Show that there are two lines through the origin that do not move under f, and that they intersect at a right angle.

4. Find the values of a and b, if the line $3x - 4y + 1 = 0$ moves to the line $3x + 2y - 1 = 0$ under the linear transformation expressed by the matrix $\begin{pmatrix} 1 & a \\ b & 5 \end{pmatrix}$.

5. Two lines $x + y = 1$ and $2x - y = 1$ move to each other under a certain linear transformation f. Find the matrix of f.

6. Show that if two arbitrary points P_1 and P_2 on a plane move to P_1' and P_2' under the linear transformation $f: (x, y) \to (ax + by, cx + dy)$, and if $P_1 P_2 = P_1' P_2'$ always holds, then $a^2 + c^2 = b^2 + d^2 = 1$ and $ab + cd = 0$.

7. Take f as the linear transformation expressed by the matrix $\begin{pmatrix} 1 & 2 \\ -2 & -5 \end{pmatrix}$. Find the equations of lines l_1 and l_2, if line l_1 moves to line l_2 under f, and l_1 and l_2 intersect at point $P(0, 1)$.

CHAPTER 4

FIGURES IN SPACE

SECTION 1.   POINTS, STRAIGHT LINES, AND PLANES IN SPACE
SECTION 2.   COORDINATES IN SPACE
SECTION 3.   VECTORS IN SPACE
SECTION 4.   EQUATIONS OF STRAIGHT LINES AND PLANES IN SPACE

We learned about vectors in a plane in Chapter 2. In this chapter, we will learn about figures and vectors in space. The general principles are similar, but by extending our viewpoint into three dimensions, we can substantially deepen our understanding of geometry.

Straight lines, planes, and spaces are also referred to as one-, two-, and three-dimensional spaces, respectively. Today, we also think in terms of more general "n–dimensional spaces." Moreover, vectors are no longer merely a geometric concept, but have been generalized to a tremendous extent; various mathematical objects are regularly described as "vector spaces." This is characteristic of the modern method of axiomatization. The mathematician responsible for the axiomatic definition of the concept of a vector space was Giuseppe Peano (1858–1932), who also gave us the axiomatic system of natural numbers.

POINTS, STRAIGHT LINES, AND PLANES IN SPACE

Straight Lines and Planes in Space

You learned in junior high school about the relative position of straight lines and planes in space. Let us summarize what you already know.

Relative Position of Two Lines

There are three cases of the relative position of two lines in space:

(1) Intersecting   (2) Parallel   (3) Skew

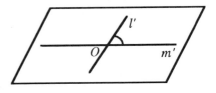

In cases (1) and (2), the two lines lie in the same plane, but in case (3) they do not lie in the same plane.

In case (3), let us draw straight lines l' and m' through an arbitrary point O parallel to l and m. Then the angle formed by l' and m' is constant, regardless of the position of point O. This angle is said to be the **angle formed by two straight lines** l and m.

As a special case of (1) and (3), when two lines l and m form a right angle, l and m are said to be **perpendicular**; we designate this relation by $l \perp m$.

Problem 1 Find the angle formed by the following pairs of segments in the cube to the right.

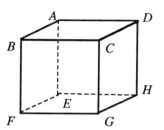

(1) BC and GH

(2) BA and FC

(3) BD and AH

Determining a Plane

In space a single plane can be determined by the information in (1) – (4), as shown in the figures below.

(1) Three points not on a single line

(2) One line and one point not on that line

(3) Two intersecting lines

(4) Two parallel lines

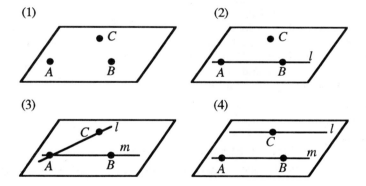

The Angle Formed by Two Planes

There are two cases of relative position of two planes in space.

(1) Intersecting   (2) Parallel

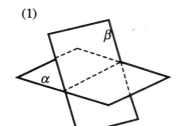
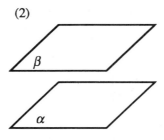

Thus, if two different planes α and β share a common point, these two planes share one line through that point. In this case, the two planes are said to be **intersecting**, and that line is called the **line of intersection**.

However, if two planes α and β share no common point, these two planes are said to be **parallel**; we designate this relation as $\alpha \parallel \beta$.

Draw lines OA and OB from point O on the line of intersection of two intersecting planes α and β perpendicular to this line of intersection. Then the measure of $\angle AOB$ is constant, regardless of the location of point O. This angle is referred to as the **angle formed by two planes**. As a special case, if $\angle AOB = \angle R$, α and β are said to be **perpendicular**; we designate this relation as $\alpha \perp \beta$.

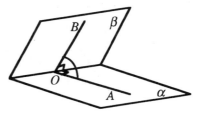

Problem 2 Find the angle formed by the following pairs of planes in the cube to the right:

(1) plane $AEGC$ and plane $AEHD$

(2) plane $ABCD$ and plane $BFHD$.

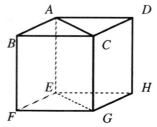

4 FIGURES IN SPACE

Straight Lines and Planes

There are three cases of the relative position of a straight line and a plane in space.

(1) Intersecting   (2) Parallel   (3) The line is included in the plane

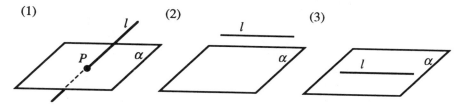

Thus, if straight line l and plane α have only one common point P, l and α are said to be **intersecting**, and that point P is referred to as the **intersection** of l and α.

However, if l and α have no common points, l and α are said to be **parallel**; we designate this relation as $l \parallel \alpha$.

If l and α have two points in common, then l is included in α.

 ## Straight Lines Perpendicular to a Plane

If a straight line l is perpendicular to all lines in plane α, l and α are said to be **perpendicular**; we designate this relation as $l \perp \alpha$.

> If straight line l is perpendicular to two intersecting lines a and b in plane α, then $l \perp \alpha$.

[Proof]   Take c as an arbitrary line in plane α, and prove that $l \perp c$. We should assume that l and c pass through the intersection O of a and b.

As in the figure to the right, draw line m such that it does not pass through O in plane α, and take A, B, and C as the points at which it intersects a, b, and c, respectively.

Take two points P and Q symmetric to each other with respect to point O, and connect them to the other points as in the figure to the right. Then

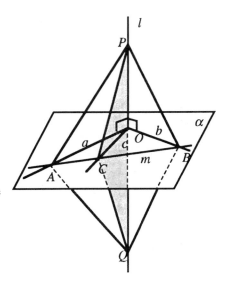

$$AP = AQ \qquad (1)$$

$$BP = BQ \qquad (2)$$

From (1) and (2), $\qquad \triangle PAB \equiv \triangle QAB.$

Therefore, $\qquad \angle PAB = \angle QAB. \qquad (3)$

From (1) and (3), $\qquad \triangle APC \equiv \triangle AQC.$

Thus, $\qquad PC = QC.$

Since O is the midpoint of the base of isosceles triangle CPQ,

$$PQ \perp CO \quad \text{or} \quad l \perp c.$$

Theorem of Three Perpendiculars

The following theorem is called the **theorem of three perpendiculars**.

> **Theorem of Three Perpendiculars**
>
> Draw line PA from point P, not in plane α, perpendicular to one line a in α, and draw line b through A perpendicular to a in α. In this situation, if we draw line PQ from P perpendicular to b, then $PQ \perp \alpha$.

[Proof]   Since

$$a \perp AP,\ a \perp AQ.$$

Line a is perpendicular to plane β determined by AP and AQ.

Therefore, $a \perp PQ.$

However, $b \perp PQ.$

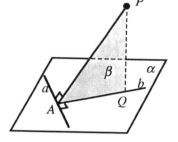

Thus, PQ is perpendicular to plane α determined by a and b.

Problem 1   Prove that if we draw line PA from point P, not in plane α, to one line a in α, and draw line PQ from P perpendicular to α, then $AQ \perp a$.

Problem 2   Prove that if we draw line PQ from point P, not in plane α, to α, and draw line QA from Q perpendicular to one line a in α, then $PA \perp a$.

Note: Problem 1 and Problem 2, taken together, make up the theorem of three perpendiculars.

Exercises

1. Given two different lines l and m and three planes α, β and γ. Mark the following statements with an O in the blanks if they hold and with an X if they do not hold.

   (1) $l \parallel \alpha$, $m \parallel \alpha$ $\Rightarrow$ $l \parallel m$ [ ]

   (2) $l \perp \alpha$, $m \perp \alpha$ $\Rightarrow$ $l \parallel m$ [ ]

   (3) $l \perp \alpha$, $m \parallel \alpha$ $\Rightarrow$ $l \perp m$ [ ]

   (4) $\alpha \perp \gamma$, $\beta \perp \gamma$ $\Rightarrow$ $\alpha \parallel \beta$ [ ]

   (5) $\alpha \parallel \gamma$, $\beta \perp \gamma$ $\Rightarrow$ $\alpha \perp \beta$ [ ]

2. State which edges of tetrahedron $ABCD$ are skew to each other.

3. Given four points $A, B, C,$ and D. If the two lines AB and CD are skew, then lines AD and BC are also skew. Prove this statement by means of an indirect proof.

4. If two planes α and β intersect at line l, draw lines PA and PB from point P perpendicular to α and β. Prove that if we draw lines AA' and BB' from A and B perpendicular to l, then A' and B' coincide.

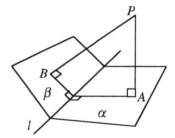

4 FIGURES IN SPACE   131

  COORDINATES IN SPACE

  Coordinates in Space

We have already learned how to represent the points on a line or in a plane with coordinates. Now let's consider coordinates in space.

Coordinates in space are defined by three orthogonal **coordinate axes** which intersect at one point O.

These axes are number lines with O as the origin, and they are called the **x-axis, y-axis**, and **z-axis**. Point O is called the **origin** of coordinates in space.

The plane determined by the x- and y-axes, the plane determined by the y- and z-axes, and the plane determined by the z- and x-axes are called the **xy-plane, yz-plane**, and **zx-plane**, respectively, and taken together they are called the **coordinate planes**.

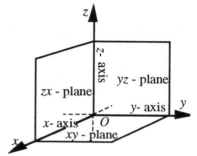

Problem 1   Which coordinate plane is perpendicular to the x-axis? Which coordinate axis intersects the zx-plane orthogonally?

Let P be a given arbitrary point.

Take A, B and C as the points at which the planes through point P parallel to the yz-, zx-, and xy-planes intersect the x-, y-, and z-axes, respectively.

When the coordinates of points A, B, and C on each coordinate axis are a, b, and c, respectively, the combination of the three numbers (a, b, c) is referred to as the **coordinates** of point P. We represent the fact that the coordinates of point P are (a, b, c) by writing $P(a, b, c)$.

The coordinates a, b, and c are referred to as the **x-coordinate**, the **y-coordinate**, and the **z-coordinate** of point P, respectively.

As we can see from the theorem of three perpendiculars, PA, PB, and PC are perpendicular to the x-, y-, and z-axes, respectively.

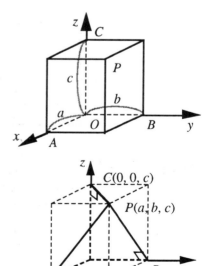

132 4 FIGURES IN SPACE

Problem 2 If we draw line PP' from the point (4, 3, 6) perpendicular to the xy-plane, find the coordinates of P'. Then do the same for the other perpendicular lines to the other coordinate planes.

Problem 3 Find the coordinates of point Q, symmetric to point $P(5, 2, -4)$ with respect to the xy-plane, and point R, symmetric to point P with respect to the origin O.

Coordinates and Translations

In general, if each point in space is translated by α units along the x-axis, by β units along the y-axis, and by γ units along the z-axis, then the point (x, y, z) moves to point

$$(x + \alpha, y + \beta, z + \gamma).$$

As a special case, the origin O moves to the point (α, β, γ).

We can formulate the following general rule for such translations.

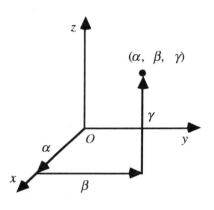

Coordinates and Translations

Under a translation that moves the origin O to the point (α, β, γ), the point that moves to (x, y, z) is the point $(x - \alpha, y - \beta, z - \gamma)$.

Problem 4 To what point do the following points move under a translation that moves the origin to the point (3, -1, 2)? What points move to the following points?

$$(2, 3, 4) \qquad (-2, -3, -4) \qquad (-3, 1, -2)$$

Problem 5 If a certain translation moves the point (4, 2, -3) to (-1, 3, 5), to what point does (1, 1, 1) move?

Interior and Exterior Dividing Points

Take (x, y, z) as the coordinates of point C, which divides line segment AB connecting the two points $A(x_1, y_1, z_1)$ and $B(x_2, y_2, z_2)$ at a ratio of $m : n$.

Draw lines AA', BB', and CC' from points A, B, and C perpendicular to the xy-plane, and then

$$AC : CB = A'C' : C'B' = m : n.$$

The coordinates of A', B' and C' are, respectively,

$$(x_1, y_1, 0), (x_2, y_2, 0), (x, y, 0).$$

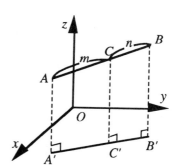

From the formula for an interior dividing point, we obtain

$$x = \frac{mx_2 + nx_1}{m + n}, \quad y = \frac{my_2 + ny_1}{m + n}.$$

Analogously, if we draw a perpendicular line to the yz-plane, we obtain

$$z = \frac{mz_2 + nz_1}{m + n}.$$

Therefore, the coordinates of point C are

$$\left(\frac{mx_2 + nx_1}{m + n}, \frac{my_2 + ny_1}{m + n}, \frac{mz_2 + nz_1}{m + n}\right).$$

Problem 6 Prove that the coordinates of the point which divides line segment AB externally at a ratio of $m : n$ are

$$\left(\frac{mx_2 - nx_1}{m - n}, \frac{my_2 - ny_1}{m - n}, \frac{mz_2 - nz_1}{m - n}\right).$$

The Distance between Two Points, and the Equation of a Sphere

If we draw line PH from point $P(x, y, z)$ in space perpendicular to the xy-plane, then the coordinates of H are $(x, y, 0)$. Therefore,

$$OH^2 = x^2 + y^2.$$

Moreover, since $\triangle POH$ is a right triangle in which H is the vertex of the right angle,

$$OP^2 = OH^2 + PH^2 = x^2 + y^2 + z^2.$$

Thus,

$$OP = \sqrt{x^2 + y^2 + z^2}.$$

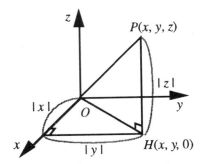

In general, given two points $P(x_1, y_1, z_1)$ and $Q(x_2, y_2, z_2)$, we can take R as the point which moves to point $Q(x_2, y_2, z_2)$ under a translation that moves the origin O to point $P(x_1, y_1, z_1)$, and then the coordinates of R are

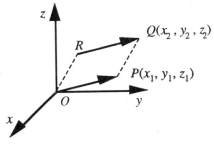

$(x_2 - x_1, y_2 - y_1, z_2 - z_1).$

Since a translation does not change the length of a line segment, $PQ = OR$. Hence, we can formulate the following generalization.

The Distance between Two Points

Take d as the distance between two points (x_1, y_1, z_1) and (x_2, y_2, z_2), and then

$$d = \sqrt{(x_2 - x_1)^2 + (y_2 - y_1)^2 + (z_2 - z_1)^2}.$$

Problem 1 Find the distance between the following pairs of points:

(1) $A(3, -2, 4)$, $B(2, 0, 3)$

(2) $A(0, 2, 3)$, $B(-4, 0, 1)$

Equation of a Sphere

The set of points at a constant distance r from a fixed point C is called a **spherical surface** with center C and radius r, or simply a **sphere**.

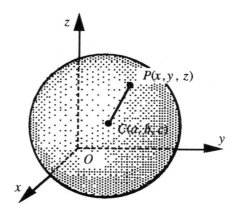

Taking (a, b, c) as the coordinates of the center C, the condition $CP = r$ such that point $P(x, y, z)$ lies on the surface can be expressed by the formula for the distance between two points as

$$(x - a)^2 + (y - b)^2 + (z - c)^2 = r^2.$$

This is the equation of a sphere.

The Equation of a Sphere

The equation of a sphere with its center at the point (a, b, c) and a radius of r is

$$(x - a)^2 + (y - b)^2 + (z - c)^2 = r^2.$$

Problem 2   Find the equations of the following spheres:

(1) a sphere with its center at (2, 0, -3) and a radius of 6;

(2) a sphere with its center at (4, -3, 5) and tangent to the yz-plane.

Demonstration   Show that $x^2 + y^2 + z^2 - 6x + 8z = 0$ is the equation of a sphere, and find its center and radius.

[Solution]   If we transform the given equation, we obtain

$$(x^2 - 6x + 9) + y^2 + (z^2 + 8z + 16) = 9 + 16$$

$$(x - 3)^2 + y^2 + (z + 4)^2 = 5^2.$$

This is the equation of a sphere with its center at (3, 0, -4) and a radius of 5.

Problem 3 Find the center and radius of the sphere

$$x^2 + y^2 + z^2 - 4x + 6y + 2z = 11.$$

The interior of a sphere is the set of points whose distance from the center is less than the radius, and therefore the interior of a sphere with its center at (a, b, c) and a radius of r can be expressed as

$$\{(x, y, z) | (x - a)^2 + (y - b)^2 + (z - c)^2 < r^2\}.$$

The exterior of this sphere is expressed as

$$\{(x, y, z) | (x - a)^2 + (y - b)^2 + (z - c)^2 > r^2\}.$$

Problem 4 Express the interior and exterior of a sphere with its center at the origin and a radius of 4 in the same way as above.

Problem 5 Express the region between a sphere with its center at the point (2, 1, -3) and a radius of 1 and a sphere with its center at the same point and a radius of 4 in the same way as above.

Exercises

1. Find the coordinates of point P', which is symmetric to point $P(5, -2, 6)$ with respect to point $A(3, 2, -4)$.

2. Find the coordinates of the points on the x- and y-axes that are equidistant from two points $A(4, 5, 3)$ and $B(3, -2, 5)$.

3. Given points $A(2, 3, 4)$, $B(-3, 2, 0)$, and $C(4, -2, 5)$.

    (1) Find the coordinates of the centroid of $\triangle ABC$.

    (2) Find the coordinates of point D such that the midpoint of AD and the midpoint of BC coincide.

4. Given points $A(4, -1, 2)$ and $B(1, 1, 3)$. Find the coordinates of point C on the xy-plane such that $\triangle ABC$ is an equilateral triangle.

5. Given points $A(0, 3, 0)$, $B(0, 1, -2)$, and $C(2, 3, -2)$.

    (1) What kind of triangle is $\triangle ABC$?

    (2) Find the equation of a sphere that passes through A, B, C, and the origin.

6. Find the equations of the following spheres:

    (1) a sphere which has as a diameter a line segment connecting the two points $(-2, 1, 5)$ and $(4, -3, -1)$;

    (2) a sphere which passes through the point $(-5, 1, 4)$ and is tangent to the three coordinate planes.

7. What figure is the locus of point P? Find the equation of that figure if

    (1) point P is such that the ratio of its distance from two points $A(-1, 0, 0)$ and $B(2, 0, 0)$ is $1 : 2$;

    (2) point P is such that the sum of the squares of its distance from two points $A(1, 2, 0)$ and $B(-1, 4, 2)$ is 38.

3 VECTORS IN SPACE

1 Vectors in Space

As when dealing with vectors in a plane, when we consider the direction and length of a directed line segment AB in space and disregard its location, we refer to it as a **vector in space**, or simply **vector**, and designate it as $\overrightarrow{AB}$.

We can also designate a vector by a single letter with an arrow, such as $\vec{a} = \overrightarrow{AB}$.

The length of directed line segment AB is called the **magnitude** of vector $\overrightarrow{AB} = \vec{a}$, and is designated by $|\overrightarrow{AB}|$ or $|\vec{a}|$.

If two vectors are identical, that means their magnitude and direction are the same. Therefore, if we obtain directed line segment CD by translating directed line segment AB, then

$$\overrightarrow{AB} = \overrightarrow{CD}.$$

Just as in a plane, if $\overrightarrow{BA}$ has the opposite direction from $\overrightarrow{AB}$, we say that it is the **inverse vector** of $\overrightarrow{AB}$, and we designate it by $-\overrightarrow{AB}$; we also say that $\overrightarrow{AA}$ is a **zero vector** and designate it by $\vec{0}$.

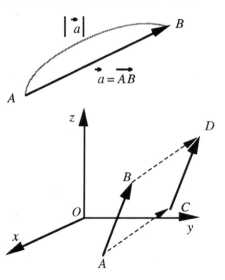

Demonstration

Given the two points $A(1, 2, 1)$ and $B(2, 1, 3)$. Taking O as the origin, find the coordinates of point P such that $\overrightarrow{OP} = \overrightarrow{AB}$. Then find the coordinates of point Q such that $\overrightarrow{OQ} = -\overrightarrow{AB}$.

4 FIGURES IN SPACE   139

[Solution]   Under the translation that moves the origin O to point $A(1, 2, 1)$, point P moves to point $B(2, 1, 3)$. Therefore, the coordinates of point P are

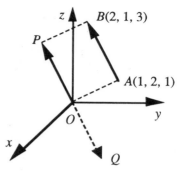

$$(2 - 1, 1 - 2, 3 - 1).$$

Thus,

$$(1, -1, 2).$$

Moreover, since

$$\overrightarrow{OQ} = -\overrightarrow{AB} = -\overrightarrow{OP}$$

point Q is symmetric to point P with respect to the origin. Therefore, the coordinates of point Q are

$$(-1, 1, -2).$$

Problem   Given points $A(2, 5, 4)$, $P(3, -2, 1)$ and $Q(4, 0, -3)$. Taking O as the origin, find the coordinates of points B and C such that $\overrightarrow{AB} = \overrightarrow{OP}$ and $\overrightarrow{AC} = \overrightarrow{OQ}$.

   Operations Involving Vectors

Given two vectors in space $\vec{a}$ and $\vec{b}$, their sum $\vec{a} + \vec{b}$ and difference $\vec{a} - \vec{b}$ are defined just as in the case of vectors in a plane, as shown in the following figures.

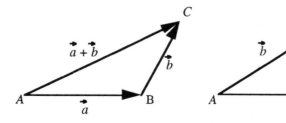

The definition of the product $m\vec{a}$ of vector $\vec{a}$ and a real number m is the same as with vectors in a plane.

Thus, $|m\vec{a}| = |m||\vec{a}|$.

If $\vec{a}$ is not $\vec{0}$, then

    for $m > 0$, $m\vec{a}$ and $\vec{a}$ have the same direction;

    for $m < 0$, $m\vec{a}$ has the opposite direction from $\vec{a}$;

    for $m = 0$, $m\vec{a}$ is the zero vector $\vec{0}$.

If $\vec{a}$ is $\vec{0}$, then

$$m\vec{a} = \vec{0}.$$

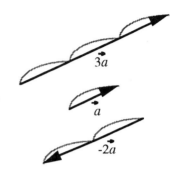

From the definition above, we can formulate the following generalization.

For $\vec{a} \neq \vec{0}$ and $\vec{b} \neq \vec{0}$,

$$\vec{a} \parallel \vec{b} \iff \vec{a} = m\vec{b}.$$

A vector with a magnitude of 1 is called a **unit vector**.

Taking $\vec{a}$ as any arbitrary vector not equal to 0, then

$$\vec{e} = \frac{\vec{a}}{|\vec{a}|}$$

is a unit vector with the same direction as $\vec{a}$.

Analogous to operations with vectors in a plane, addition of vectors in space and multiplication of vectors in space by real numbers have the following fundamental properties.

$$\vec{a} + \vec{b} = \vec{b} + \vec{a}$$

$$(\vec{a} + \vec{b}) + \vec{c} = \vec{a} + (\vec{b} + \vec{c})$$

$$(mn)\vec{a} = m(n\vec{a})$$

$$(m + n)\vec{a} = m\vec{a} + n\vec{a}$$

$$m(\vec{a} + \vec{b}) = m\vec{a} + m\vec{b}$$

  Problem Express $\vec{x}$, which satisfies the following equations, in terms of $\vec{a}$ and $\vec{b}$:

(1) $\quad 3(\vec{a} + \vec{x}) = 2(\vec{x} - 3\vec{b})$

(2) $\quad 2\vec{b} - 7\vec{x} = 3(4\vec{a} + \vec{b}) + 4(2\vec{a} + \vec{b} - 3\vec{x})$

③ Components of Vectors

In space, we will designate the unit vectors with the same direction as the positive direction of each coordinate axis by $\vec{e}_1$, $\vec{e}_2$, and $\vec{e}_3$, as in the figure to the right, and call them the **basic vectors** of the x-axis, the y-axis, and the z-axis.

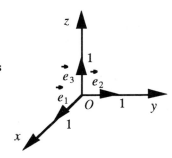

Given one vector $\vec{a}$ in space, let us take point P satisfying

$$\vec{a} = \overrightarrow{OP}.$$

Let the coordinates of point P be (a_1, a_2, a_3), and take points $P_1(a_1, 0, 0)$, $P_2(0, a_2, 0)$, and $P_3(0, 0, a_3)$ on the x-, y-, and z-axes, respectively.

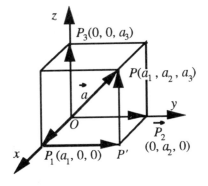

Draw line PP' from point P perpendicular to the xy-plane, and then it is clear from the figure to the right that

$$\overrightarrow{OP} = \overrightarrow{OP_1} + \overrightarrow{P_1P'} + \overrightarrow{P'P}$$

$$\overrightarrow{P_1P'} = \overrightarrow{OP_2}, \quad \overrightarrow{P'P} = \overrightarrow{OP_3}.$$

Therefore,

$$\vec{a} = \overrightarrow{OP} = \overrightarrow{OP_1} + \overrightarrow{OP_2} + \overrightarrow{OP_3}. \qquad (1)$$

From the definition of a vector multiplied by a real number,

$$\overrightarrow{OP_1} = a_1\vec{e_1}, \quad \overrightarrow{OP_2} = a_2\vec{e_2}, \quad \overrightarrow{OP_3} = a_3\vec{e_3}.$$

Substituting these expressions into (1), we obtain

$$\vec{a} = a_1\vec{e_1} + a_2\vec{e_2} + a_3\vec{e_3}. \qquad (2)$$

We refer to a_1, a_2, and a_3 in (2) as the **x-component**, the **y-component**, and the **z-component**, respectively.

From the formula for the distance between two points, the magnitude of vector $\vec{a}$ is

$$|\vec{a}| = |\overrightarrow{OP}| = \sqrt{a_1^2 + a_2^2 + a_3^2}$$

4 FIGURES IN SPACE 143

Demonstration Given two points $A(2, 1, 3)$ and $B(3, 5, 2)$. Express vector $\overrightarrow{AB}$ in terms of the basic vectors.

[Solution] Take point P as the point that moves to point B under a translation that moves the origin to point A. The coordinates of P are

$$(3 - 2, 5 - 1, 2 - 3) \text{ or } (1, 4, -1).$$

Therefore, expressing $\overrightarrow{AB} = \overrightarrow{OP}$ in terms of the basic vectors, we obtain

$$\overrightarrow{AB} = \vec{e}_1 + 4\vec{e}_2 - \vec{e}_3.$$

Problem 1 Given three points $A(-2, 3, -1)$, $B(4, -7, 5)$ and $C(3, -6, 0)$. Express $\overrightarrow{AB}$, $\overrightarrow{BC}$, and $\overrightarrow{CA}$ in terms of the basic vectors.

If the x-, y-, and z-components of vector $\vec{a}$ are a_1, a_2, and a_3, respectively, that is,

$$\vec{a} = a_1\vec{e}_1 + a_2\vec{e}_2 + a_3\vec{e}_3$$

vector $\vec{a}$ is designated by (a_1, a_2, a_3). Thus,

$$\vec{a} = (a_1, a_2, a_3).$$

This is called the **component representation** of vector $\vec{a}$.

$\vec{a} = a_1\vec{e}_1 + a_2\vec{e}_2 + a_3\vec{e}_3$	Representation in terms of the basic vectors
$\vec{a} = (a_1, a_2, a_3)$	Component representation

As special cases,

$$\vec{e}_1 = (1, 0, 0), \quad \vec{e}_2 = (0, 1, 0), \quad \vec{e}_3 = (0, 0, 1)$$

$$\vec{0} = (0, 0, 0).$$

Moreover,

$$(a_1, a_2, a_3) = (b_1, b_2, b_3) \leftrightarrow a_1 = b_1, a_2 = b_2, a_3 = b_3.$$

Problem 2 Given two points $A(4, 3, -5)$ and $B(-2, 4, 3)$. Find the component representation of $\vec{AB}$. Then find the coordinates of point D which satisfy $\vec{AB} = \vec{CD}$ for point $C(3, -2, 0)$.

Calculation by Means of Components

When we express vectors by component representations, we obtain the following formulas, analogous to those which apply for vectors in a plane.

Calculation by Means of Components

(1) $(a_1, a_2, a_3) + (b_1, b_2, b_3) = (a_1 + b_1, a_2 + b_2, a_3 + b_3)$

(2) $(a_1, a_2, a_3) - (b_1, b_2, b_3) = (a_1 - b_1, a_2 - b_2, a_3 - b_3)$

(3) $m(a_1, a_2, a_3) = (ma_1, ma_2, ma_3)$

Problem 3 Given $\vec{a} = (2, -3, 0)$, $\vec{b} = (-3, 2, 5)$, and $\vec{c} = (6, 0, -3)$. Find the component representations of vectors $\vec{u}$ and $\vec{v}$:

(1) $\vec{u} = 3\vec{a} - 4\vec{b}$ (2) $\vec{v} = 2(\vec{a} + \vec{b}) - 3(\vec{b} - 2\vec{c})$

Problem 4 If none of the components of vectors $\vec{a} = (a_1, a_2, a_3)$ and $\vec{b} = (b_1, b_2, b_3)$ are equal to 0, prove that

$$\vec{a} \parallel \vec{b} \quad \Leftrightarrow \quad \frac{a_1}{b_1} = \frac{a_2}{b_2} = \frac{a_3}{b_3}.$$

4 FIGURES IN SPACE

The following formula holds for the magnitude of a vector.

The Magnitude of a Vector

If $\vec{a} = (a_1, a_2, a_3)$, then $|\vec{a}| = \sqrt{a_1^2 + a_2^2 + a_3^2}$.

Problem 5 Find the magnitude of the following vectors:

$$\vec{a} = (3, -4, 5), \quad \vec{b} = (3 - \sqrt{2}, \sqrt{10}, 3 + \sqrt{2}).$$

Given vector $\vec{a} = (a_1, a_2, a_3)$, take point P satisfying $\vec{a} = \overrightarrow{OP}$, and take α, β, and γ as the angles formed by OP and the positive directions of the x-, y-, and z-axes, respectively, as in the figure to the right. Then

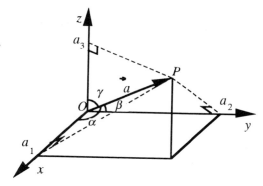

$a_1 = |\overrightarrow{OP}| \cos \alpha,$

$a_2 = |\overrightarrow{OP}| \cos \beta,$

$a_3 = |\overrightarrow{OP}| \cos \gamma.$

Therefore,

$$|\overrightarrow{OP}|^2 = a_1^2 + a_2^2 + a_3^2$$
$$= |\overrightarrow{OP}|^2 (\cos^2 \alpha + \cos^2 \beta + \cos^2 \gamma).$$

Thus,

$$\cos^2 \alpha + \cos^2 \beta + \cos^2 \gamma = 1.$$

These values $\cos \alpha$, $\cos \beta$, and $\cos \gamma$ are referred to as the **direction cosines** of vector $\vec{a} = \overrightarrow{OP}$. The vector $(\cos \alpha, \cos \beta, \cos \gamma)$ is a unit vector with the same direction as $\vec{a}$.

Problem 6 Given that the coordinates of point P are $(-2, 6, 9)$, find the direction cosines of $\overrightarrow{OP}$.

4 The Inner Product of Vectors

As with vectors in a plane, we can define the inner product $\vec{a} \cdot \vec{b}$ of two vectors in space ($\vec{a}$ and $\vec{b}$ not equal to $\vec{0}$) in terms of the angle θ formed by $\vec{a}$ and $\vec{b}$:

$$\vec{a} \cdot \vec{b} = |\vec{a}||\vec{b}|\cos\theta.$$

If $\vec{a} = \vec{0}$ or $\vec{b} = \vec{0}$, we define $\vec{a} \cdot \vec{b} = 0$.

It is clear that

$$\vec{a} \cdot \vec{b} = \vec{b} \cdot \vec{a}$$

$$|\vec{a} \cdot \vec{b}| \leq |\vec{a}||\vec{b}|$$

$$\vec{a} \cdot \vec{a} = |\vec{a}|^2.$$

Furthermore,

$$\vec{a} \perp \vec{b} \leftrightarrow \vec{a} \cdot \vec{b} = 0.$$

Example Since the basic vectors $\vec{e}_1$, $\vec{e}_2$, and $\vec{e}_3$ are unit vectors that intersect each other orthogonally,

$$\vec{e}_1 \cdot \vec{e}_1 = \vec{e}_2 \cdot \vec{e}_2 = \vec{e}_3 \cdot \vec{e}_3 = 1$$

$$\vec{e}_1 \cdot \vec{e}_2 = \vec{e}_1 \cdot \vec{e}_3 = \vec{e}_2 \cdot \vec{e}_3 = 0.$$

4 FIGURES IN SPACE

Problem 1 Given a cube of edge a, as in the figure to the right. Find the following inner products of the vectors of this cube:

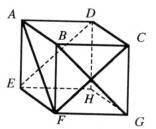

(1) $\overrightarrow{AB} \cdot \overrightarrow{AF}$ (2) $\overrightarrow{AF} \cdot \overrightarrow{BG}$

(3) $\overrightarrow{BG} \cdot \overrightarrow{DE}$ (4) $\overrightarrow{DE} \cdot \overrightarrow{FC}$

Inner Products and Component Representation

For vectors in space, the following generalizations hold just as in the case of vectors in a plane.

Inner Products and Components

Given $\vec{a} = (a_1, a_2, a_3)$ and $\vec{b} = (b_1, b_2, b_3)$,

$$\vec{a} \cdot \vec{b} = a_1 b_1 + a_2 b_2 + a_3 b_3.$$

As a special case, $\vec{a} \perp \vec{b} \iff a_1 b_1 + a_2 b_2 + a_3 b_3 = 0.$

If we take θ as the angle formed by $\vec{a} = (a_1, a_2, a_3)$ and $\vec{b} = (b_1, b_2, b_3)$, then

$$\cos\theta = \frac{\vec{a} \cdot \vec{b}}{|\vec{a}||\vec{b}|} = \frac{a_1 b_1 + a_2 b_2 + a_3 b_3}{\sqrt{a_1^2 + a_2^2 + a_3^2}\sqrt{b_1^2 + b_2^2 + b_3^2}}.$$

Problem 2 Find the inner products of the following pairs of vectors $\vec{a}$ and $\vec{b}$:

(1) $\vec{a} = (-2, 2, 3)$, $\vec{b} = (4, 5, 6)$

(2) $\vec{a} = 4\vec{e}_1 + 3\vec{e}_2 - \vec{e}_3$, $\vec{b} = -2\vec{e}_1 + \vec{e}_2 + 3\vec{e}_3$

148　4　FIGURES IN SPACE

Problem 3 Find the angle θ formed by the following pairs of vectors $\vec{a}$ and $\vec{b}$:

(1) $\vec{a} = (-1, 0, 1)$,　　$\vec{b} = (-1, 2, 2)$

(2) $\vec{a} = (-3, 2, 1)$,　　$\vec{b} = (2, 1, 4)$

The formulas on page 53 for the inner product of vectors in space hold as given. Thus,

$$\vec{a} \cdot (\vec{b} + \vec{c}) = \vec{a} \cdot \vec{b} + \vec{a} \cdot \vec{c}$$

$$(\vec{a} + \vec{b}) \cdot \vec{c} = \vec{a} \cdot \vec{c} + \vec{b} \cdot \vec{c}$$

$$\vec{a} \cdot (m\vec{b}) = m(\vec{a} \cdot \vec{b})$$

$$(m\vec{a}) \cdot \vec{b} = m(\vec{a} \cdot \vec{b}).$$

Problem 4 Given $\vec{a} = (2, -1, 4)$ and $\vec{b} = (-4, 5, 3)$, define the value of real number k such that $\vec{a} - k\vec{b}$ and $\vec{a}$ are perpendicular.

Exercises

1. Given $\vec{a} = (4, -2, 5)$ and $\vec{b} = (7, 9, -8)$, find vector $\vec{x}$ satisfying the following relations:

    (1) $2\vec{a} + \vec{x} = 3\vec{b}$　　　　　　　(2) $4\vec{x} - \vec{a} = 3\vec{a} - 4\vec{b} + 2\vec{x}$

2. Given $\vec{a} = (1, 1, 0)$, $\vec{b} = (1, 0, 1)$, $\vec{c} = (0, 1, 1)$, and $\vec{p} = (5, 6, 7)$, express $\vec{p}$ in the form $\vec{p} = l\vec{a} + m\vec{b} + n\vec{c}$.

3. Find the direction cosines of the following vectors:

    (1) $\vec{a} = (3, -4, 5)$　　　　　　　　(2) $\vec{b} = (-3, 2, 2\sqrt{3})$

4. Given vectors $\vec{a} = (2, -1, -5)$, $\vec{b} = (3x, 6, 4y - 2)$, and $\vec{c} = (z - 1, 2, z + 1)$.

   (1) Specify the values of x and y such that $\vec{a} \parallel \vec{b}$.

   (2) Specify the value of z such that $\vec{a} \perp \vec{c}$.

5. Specify the values of x, y, and z such that three vectors $\vec{a} = (x, 4, 6)$, $\vec{b} = (2, y, 6)$, and $\vec{c} = (2, 4, z)$ are perpendicular to each other.

6. Given $\vec{a} = (1, 2, -3)$ and $\vec{b} = (2, -1, -2)$. If vector $\vec{x} = (x_1, x_2, x_3)$ intersects both $\vec{a}$ and $\vec{b}$ orthogonally, find $x_1 : x_2 : x_3$. Here, $\vec{x} \neq \vec{0}$.

7. Given $\vec{a} = (2, -2, 1)$ and $\vec{b} = (2, 3, -4)$.

   (1) Let $\vec{b} - k\vec{a} = \vec{c}$, where k is a real number. If $\vec{c}$ and $\vec{a}$ intersect orthogonally, find the values of k and $\vec{c}$.

   (2) Find vector $\vec{a}$, which has a magnitude of 3 and intersects both $\vec{a}$ and $\vec{b}$ orthogonally.

4 EQUATIONS OF STRAIGHT LINES AND PLANES IN SPACE

1 Position Vectors

A vector in space

$$\overrightarrow{OP} = \vec{p}$$

is determined for an arbitrary point P if we determine the fixed point O. In this situation, $\vec{p}$ is referred to as the **position vector** of point P with point O as the initial point.

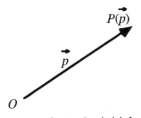

We designate that the position vector of point P is $\vec{p}$ by $P(\vec{p})$.

For vectors in space, given two points $A(\vec{a})$ and $B(\vec{b})$, the position vectors $\vec{c}$ and $\vec{d}$ of point C, which divides line segment AB internally at a ratio of $m:n$, and point D, which divides line segment AB externally at a ratio of $m:n$, are given by the following formulas:

$$\vec{c} = \frac{m\vec{b} + n\vec{a}}{m + n}, \quad \vec{d} = \frac{m\vec{b} - n\vec{a}}{m - n}.$$

Problem 1 Find the position vector $\vec{g}$ of the centroid G of $\triangle ABC$, in which the three points $A(\vec{a})$, $B(\vec{b})$, and $C(\vec{c})$ are the vertices.

Position Vectors and Coordinates

Given a set of coordinate axes, we usually take the origin as the initial point of a position vector.

In this case if we take $\vec{p}$ as the position vector of point $P(x, y, z)$, then you can see from the figure to the right that

$$\vec{p} = x\vec{e_1} + y\vec{e_2} + z\vec{e_3}.$$

As a component representation, we have $\vec{p} = (x, y, z)$, so if we take the origin as the initial point of the position vector, the coordinates of P automatically give the component representation of P.

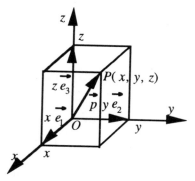

Problem 2 Find the coordinates of the centroid of a triangle in which the vertices are the three points $A(6, -2, 3)$, $B(-1, 6, 3)$, and $C(1, 5, -3)$.

Figures and Position Vectors

Demonstration Given four points $O, A, B,$ and C in space, prove that if $OB \perp CA$, and $OC \perp AB$, then $OA \perp BC$.

[Proof] Take point O as the origin, and take $\vec{a}, \vec{b},$ and $\vec{c}$ as the position vectors of the three points $A, B,$ and C.

Since $OB \perp CA$,

$$\vec{b} \cdot (\vec{a} - \vec{c}) = 0.$$

Therefore,

$$\vec{b} \cdot \vec{a} = \vec{b} \cdot \vec{c}. \quad (1)$$

Since $OC \perp AB$,

$$\vec{c} \cdot (\vec{b} - \vec{a}) = 0.$$

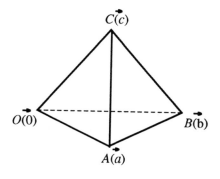

Therefore,

$$\vec{c} \cdot \vec{b} = \vec{c} \cdot \vec{a}. \qquad (2)$$

From (1) and (2),

$$\vec{b} \cdot \vec{a} = \vec{c} \cdot \vec{a}.$$

Therefore,

$$(\vec{b} - \vec{c}) \cdot \vec{a} = 0.$$

Thus,

$$\overrightarrow{CB} \cdot \overrightarrow{OA} = 0.$$

Accordingly,

$$OA \perp BC.$$

Problem 3 Prove that if four points A, B, C, and D in space satisfy $AB^2 + CD^2 = AC^2 + BD^2$, then $AD \perp BC$.

Equation of a Straight Line in Space

Given one point $P_0(\vec{p_0})$ in space and a vector $\vec{u}$ not equal to $\vec{0}$, let's consider the equation of a line through P_0 and parallel to $\vec{u}$.

Analogous to the case we examined on page 60, let us take $P(\vec{p})$ as a point moving along l, and then

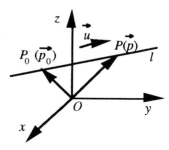

$$\overrightarrow{P_0P} \parallel \vec{u}, \quad \overrightarrow{P_0P} = \vec{p} - \vec{p_0}$$

Therefore, with a real number t, we can write

$$\vec{p} - \vec{p_0} = t\vec{u} .$$

Thus,

$$\vec{p} = \vec{p_0} + t\vec{u} . \tag{1}$$

Taking

$$\vec{p} = (x, y, z), \ \vec{p_0} = (x_0, y_0, z_0), \text{ and } \vec{u} = (a, b, c)$$

and substituting these components into (1), we obtain the following equalities with respect to each component:

$$\begin{cases} x = x_0 + ta \\ y = y_0 + tb \\ z = z_0 + tc \end{cases} . \tag{2}$$

The equalities in (1) and (2) are referred to as the **parametric representation** of line l, and t is called the **parameter**. Moreover, $\vec{u} = (a, b, c)$ is called the **direction vector** of line l.

154 4 FIGURES IN SPACE

As long as a, b, and c are not equal to 0, eliminating t from (2) gives us the following result.

A Straight Line through One Point

The equation of a line through the point $P_0(x_0, y_0, z_0)$ parallel to the vector $\vec{u} = (a, b, c)$ is

$$\frac{x - x_0}{a} = \frac{y - y_0}{b} = \frac{z - z_0}{c}.$$

If there is a 0 among a, b, and c, for example $a \neq 0$, $b \neq 0$, and $c = 0$, from (2) we get

$$\frac{x - x_0}{a} = \frac{y - y_0}{b}, \quad z = z_0.$$

This equation represents a line parallel to the xy-plane.

For example, for the case when $a \neq 0$, $b = 0$, and $c = 0$, (2) gives us

$$y = y_0, \quad z = z_0.$$

This equation represents a line parallel to the x-axis.

Problem 1 Find the equation of a line which passes through the point (3, 4, -2) and has the following direction vectors:

(1) (-2, 1, 3)   (2) (5, 1, 4)

(3) (3, 4, 0)   (4) (2, 0, 0)

Problem 2 Find the equation of a line through the point (5, -2, 3) and parallel to the line $x - 2 = \dfrac{y - 5}{3} = \dfrac{4 - z}{6}$.

Next, let's consider the equation of a line through two points

$P_0(x_0, y_0, z_0)$ and $P_1(x_1, y_1, z_1)$.

For the direction vector of this line, we can take

$$\overrightarrow{P_0 P_1} = (x_1 - x_0, y_1 - y_0, z_1 - z_0).$$

Therefore, when

$$x_1 \neq x_0, \ y_1 \neq y_0, \ z_1 \neq z_0,$$

the equation of the line takes the following form.

A Straight Line through Two Points

The equation of a straight line through two points (x_0, y_0, z_0) and (x_1, y_1, z_1) is

$$\frac{x - x_0}{x_1 - x_0} = \frac{y - y_0}{y_1 - y_0} = \frac{z - z_0}{z_1 - z_0}.$$

As another example, for the case when

$$x_1 \neq x_0, \ y_1 \neq y_0, \ z_1 = z_0$$

the equation of the line above is

$$\frac{x - x_0}{x_1 - x_0} = \frac{y - y_0}{y_1 - y_0}, \ z = z_0.$$

Problem 3 Find the equations of the lines through the following pairs of points:

(1) $(-1, 2, 3)$, $(4, -5, 6)$ (2) $(-3, 0, 2)$, $(4, 7, -5)$

(3) $(3, 4, -2)$, $(5, 3, -2)$ (4) $(-2, 5, 3)$, $(-2, 0, 3)$

Problem 4 Find the equation of a line through the origin and point $P(x_1, y_1, z_1)$, which is different from the origin.

Demonstration Take l as the line $\dfrac{x}{3} = y + 6 = \dfrac{z-2}{2}$. Find the coordinates of point Q on l such that the distance PQ from the point $P(3, -1, 2)$ is a minimum.

[Solution] Take (x, y, z) as the coordinates of point Q, and then

$$PQ^2 = (x-3)^2 + (y+1)^2 + (z-2)^2. \quad (1)$$

Since point Q is on line l,

$$\begin{cases} x = 3t \\ y = t - 6 \\ z = 2t + 2 \end{cases} \quad (2)$$

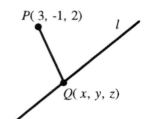

Substituting (2) into (1), we obtain

$$\begin{aligned} PQ^2 &= (3t-3)^2 + (t-5)^2 + 4t^2 \\ &= 14t^2 - 28t + 34 \\ &= 14(t-1)^2 + 20. \end{aligned}$$

Therefore, PQ^2 is a minimum for $t = 1$. Thus, the coordinates of point Q when PQ is a minimum are $(3, -5, 4)$.

Problem 5 Check that when PQ is a minimum, PQ intersects line l orthogonally.

Equation of a Plane

Given one point $P_0(\vec{p_0})$ in space and a vector $\vec{n}$ not equal to $\vec{0}$, let's find the equation of a plane α which passes through P_0 perpendicular to $\vec{n}$.

Take an arbitrary point $P(\vec{p})$ on α,

$$\overrightarrow{P_0P} = \vec{p} - \vec{p_0}, \quad \overrightarrow{P_0P} \perp \vec{n}.$$

Therefore,

$$\vec{n} \cdot (\vec{p} - \vec{p_0}) = 0. \tag{1}$$

This is the vector equation of plane α.

If we take

$$\vec{p} = (x, y, z), \ \vec{p_0} = (x_0, y_0, z_0), \text{ and } \vec{n} = (a, b, c),$$

then

$$\vec{p} - \vec{p_0} = (x - x_0, y - y_0, z - z_0)$$

$$\vec{n} \cdot (\vec{p} - \vec{p_0}) = a(x - x_0) + b(y - y_0) + c(z - z_0).$$

Therefore, (1) can be expressed in the form

$$a(x - x_0) + b(y - y_0) + c(z - z_0) = 0.$$

A Plane through One Point

The equation of a plane which passes through a point $P_0(x_0, y_0, z_0)$ perpendicular to vector $\vec{n} = (a, b, c)$ is

$$a(x - x_0) + b(y - y_0) + c(z - z_0) = 0. \tag{2}$$

Vector $\vec{n} = (a, b, c)$ is referred to as the **normal vector** of this plane.

Problem 1 Find the equation of a plane through the point (2, 4, 5) with the normal vector (3, -1, -2).

Problem 2 Given two points $A(3, 2, 5)$ and $B(4, -2, 1)$. Find the equation of a plane through point A perpendicular to $\overrightarrow{AB}$.

Problem 3 Find the equations of two planes which pass through the point (-5, -2, 3) parallel to the xy- and yz-planes.

If we set $d = -ax_0 - by_0 - cz_0$ in equation (2) above, (2) takes on the form

$$ax + by + cz + d = 0.$$

This is the form of a linear equation in $x, y,$ and z.

Conversely, for $(a, b, c) \neq \vec{0}$, a linear equation in $x, y,$ and z

$$ax + by + cz + d = 0 \qquad (3)$$

expresses a plane.

We can see this because, for example, for $a \neq 0$, we can transform (3) into

$$a(x + \frac{d}{a}) + by + cz = 0. \qquad (4)$$

Taking

$$x_0 = -\frac{d}{a}, \quad y_0 = 0, \quad z_0 = 0$$

(4) then takes on the form of (2) above. Therefore, this equation represents a plane through the point (x_0, y_0, z_0) and perpendicular to the vector (a, b, c).

Linear Equations and Planes

A linear equation $ax + by + cz + d = 0$ in $x, y,$ and z expresses a plane.

Problem 4 What kind of plane is expressed by the linear equation
$ax + by + cz = 0$?

Demonstration 1 Find the equation of a plane through the three points $A(4, 0, 0)$, $B(2, 0, 3)$, and $C(0, -1, 4)$.

[Solution] Take the equation of the plane we want to find as

$$ax + by + cz + d = 0. \qquad (1)$$

Substituting in the coordinates of A, B, and C, we obtain

$$\begin{cases} 4a + d = 0 & (2) \\ 2a + 3c + d = 0 & (3) \\ -b + 4c + d = 0 . & (4) \end{cases}$$

From (2), (3), and (4), we can express a, b, and c in terms of d, and then

$$a = -\frac{d}{4}, \quad b = \frac{d}{3}, \quad c = -\frac{d}{6}.$$

Substituting these expressions into (1), we get

$$-\frac{d}{4}x + \frac{d}{3}y - \frac{d}{6}z + d = 0.$$

Therefore,

$$3x - 4y + 2z - 12 = 0.$$

Problem 5 Find the equation of a plane through the origin and the two points $(1, 0, -1)$ and $(0, 2, 3)$.

160 4 FIGURES IN SPACE

Demonstration 2   Find the equation of a plane through the point (3, 5, -2) and parallel to the plane $2x - y + 3z = 0$.

[Solution]   The given plane is perpendicular to the vector (2, -1, 3). Therefore, it is sufficient to find the equation of a plane through the point (3, 5, -2) with the normal vector (2, -1, 3).

Thus,

$$2(x - 3) - (y - 5) + 3(z + 2) = 0.$$

Therefore,

$$2x - y + 3z + 5 = 0.$$

Problem 6   Find the equation of a plane through the point (2, 0, -4) and parallel to the plane $3x + 4y - z = 5$.

The Distance between a Plane and a Point

Just as we can find the distance between a point and a line in a plane, the following formula enables us to find the distance between a point and a plane in space.

The Distance between a Point and a Plane

Take k as the distance between the point (x_1, y_1, z_1) and the plane $ax + by + cz + d = 0$. Then

$$k = \frac{|ax_1 + by_1 + cz_1 + d|}{\sqrt{a^2 + b^2 + c^2}}.$$

Problem 7   Find the distance:

(1) between the point (3, 1, 4) and the plane $x - 2y + 2z + 3 = 0$;

(2) between the origin and the plane $3x + 2y - 4z + 5 = 0$.

4 FIGURES IN SPACE

Demonstration 3 Express the equation of the intersection l of the two planes

$$\begin{cases} 2x - y + z = 3 & (1) \\ x + 2y - 4z = 4 & (2) \end{cases}$$

in the form

$$\frac{x-p}{a} = \frac{y-q}{b} = \frac{z}{c}.$$

[Solution] First let us express z in terms of x and y from (1) and (2).

From (1) × 2 + (2), $\quad 5x - 2z = 10$

$$z = \frac{5x - 10}{2}. \qquad (3)$$

From (1) − (2) × 2, $\quad -5y + 9z = -5$

$$z = \frac{5y - 5}{9}. \qquad (4)$$

From (3) and (4),

$$\frac{5x - 10}{2} = \frac{5y - 5}{9} = z.$$

Dividing both sides by 5, we obtain

$$\frac{x - 2}{2} = \frac{y - 1}{9} = \frac{z}{5}.$$

Problem 8 In the above Demonstration, the line of intersection l is perpendicular to the normal vectors of the two planes. Use this fact to find the direction vector of l.

162 4 FIGURES IN SPACE

Spheres and Vectors

Just as we defined the vector equation of a circle in a plane, we can consider the vector equation of a sphere in space.

If we take $P(\vec{p})$ as an arbitrary point on a sphere with point $C(\vec{c})$ as its center and a radius of r, then

$$|\overrightarrow{CP}| = r.$$

Thus,

$$|\vec{p} - \vec{c}| = r. \qquad (1)$$

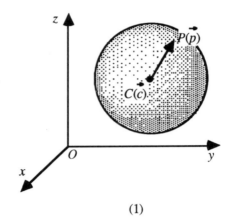

This is the vector equation of a sphere.

Squaring both sides of (1), we obtain

$$|\vec{p} - \vec{c}|^2 = r^2.$$

In this formula, if we take $\vec{p} = (x, y, z)$ and $\vec{c} = (x_0, y_0, z_0)$, then

$$(x - x_0)^2 + (y - y_0)^2 + (z - z_0)^2 = r^2.$$

This equality is identical to the equation of a sphere, which we learned on page 135.

Demonstration   On what surface does point P, the midpoint of OQ, lie, if Q is a point moving on a sphere with its center at point $A(0, 4, 0)$ and a radius of 2? O is the origin.

[Solution]   Take $\vec{a}$, $\vec{q}$, and $\vec{p}$ as the position vectors of A, Q, and P. Since point Q is a point on a sphere,

$$|\vec{q} - \vec{a}| = 2. \qquad (1)$$

Since point P is the midpoint of OQ,

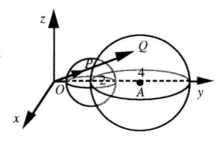

$$\vec{p} = \frac{1}{2}\vec{q} \text{ or } \vec{q} = 2\vec{p}.\tag{2}$$

Substituting (2) into (1), we obtain

$$|2\vec{p} - \vec{a}| = 2.$$

Since $|2\vec{p} - \vec{a}| = 2|\vec{p} - \frac{\vec{a}}{2}|$,

$$|\vec{p} - \frac{\vec{a}}{2}| = 1.$$

Therefore, point P lies on a sphere with its center at $(0, 2, 0)$ and a radius of 1.

Problem 1 In the above Demonstration, if we take P as a point which divides OQ externally at a ratio of $3 : 1$, on what surface does point P lie? Find the vector equation of that surface.

Problem 2 Taking $\vec{a} = (4, -6, 8)$, find the center and the radius of the sphere $|2\vec{p} - 3\vec{a}| = 6$.

Exercises

1. Given a regular tetrahedron $ABCD$ with an edge of a. Take M as the midpoint of edge CD, and then:

   (1) find the inner products $\vec{AM} \cdot \vec{CD}$ and $\vec{BM} \cdot \vec{CD}$;

   (2) prove that $AB \perp CD$.

2. (1) Find the equation of a line through the point (2, -3, 7) with a direction vector of (1, 1, -4).

   (2) Find the direction vector of the line $2x - 6 = 4 - y = z - 5$.

   (3) Find the angle formed by (1) and (2).

3. Take l as the line $\dfrac{x-1}{2} = \dfrac{2-y}{3} = z + 2$, and α as the plane $3x - y - 2z = 12$.

   (1) Find the coordinates of P, the point at which l intersects plane α.

   (2) Find the angle formed by line l and the normal vector of plane α.

4. Find the equation of the following planes:

   (1) a plane through the point (5, 3, 4) parallel to the yz-plane;

   (2) a plane through the point (3, -2, 5) perpendicular to the vector (-4, 2, -3);

   (3) a plane through the point (4, -2, 3) parallel to the plane $3x + 6y - 4z = 7$;

   (4) a plane through the three points (3, 0, 0), (0, 4, 0), and (0, 0, 5).

5. Given two planes $3x + z - 1 = 0$ and $x - \sqrt{5}y + 2z = 0$. Find the normal vector of each plane, and find the angle formed by these two planes.

6. Find the equation of a plane through the point (-2, 1, 3) and perpendicular to both of the planes $x - y + z = 0$ and $2x + 3y - z = 5$.

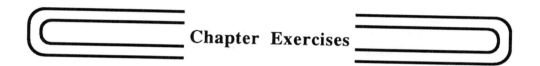

Chapter Exercises

A

1. Show that the four line segments connecting each vertex of a tetrahedron and the centroid of the opposite face intersect at the point which divides these four line segments internally at a ratio of 3 : 1.

2. A sphere with its center at (3, 7, 4) and a radius of 5 intersects the xy- and yz-planes, forming two circles. Find the center and radius of each circle.

3. (1) Specify the values of m and n such that the following three points all lie on one line:

   $A(2, 3, -4)$, $B(3, 1, -1)$, $C(m, 7, n - 1)$

   (2) Specify the value of m such that the following four points lie in one plane:

   $P(4, -2, 5)$, $Q(-3, 4, -4)$, $R(1, 2, 4)$, $S(m, 1 - m, 4)$

4. Find the equation of a line through the point (3, -1, 2) which forms angles of 60°, 45°, and 60° with the x-, y-, and z-axes, respectively.

5. Show that the two lines $\dfrac{x}{2} = \dfrac{y-2}{3} = z + 4$ and $\dfrac{x-1}{2} = \dfrac{y}{3} = z$ are parallel, and find the equation of the plane determined by these two lines.

6. Show that the two lines $\dfrac{x-1}{3} = \dfrac{y-3}{4} = \dfrac{z+2}{5}$ and $\dfrac{x-1}{2} = \dfrac{y-3}{5} = \dfrac{z+2}{3}$ intersect at a single point, and find the equation of the plane determined by these two lines.

7. Find the volume of a solid body bordered by the xy-plane, yz-plane, and zx-plane and the plane $6x + 4y + 3z = 12$.

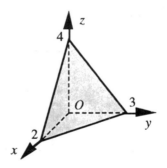

8. Take α as the plane $4x + 3y + 5z = 50$.

   (1) Find the distance between plane α and the origin.
   (2) Find the coordinates of a point symmetric to the origin with respect to plane α.

B

1. If the magnitudes of two vectors $t\vec{a} + \vec{b}$ and $\vec{a} - t\vec{b}$ are equal, find the angle formed by vectors $\vec{a}$ and $\vec{b}$, provided that $|\vec{a}| = |\vec{b}| \neq 0$ and $t \neq 0$.

2. (1) Taking $\vec{a} = (1, 1, \sqrt{2})$ and $\vec{b} = (1, 2, 3)$, find the value of t such that $|\vec{b} - t\vec{a}|$ is a minimum.

   (2) Take t_0 as the value found in (1), and then show that $\vec{b} - t_0\vec{a}$ is perpendicular to $\vec{a}$.

3. What figure is the locus of the point P, equidistant from the three points $A(4, 3, -2)$, $B(4, -3, 6)$, and $C(-4, 5, 6)$? Find the equation of that figure.

4. Take A as the point $(-2, 1, 3)$ and α as the plane $3x + y - 2z = 17$.

   (1) Find the coordinates of point H, at which a perpendicular line through point A to plane α intersects plane α.

   (2) Find the coordinates of the point B, symmetric to the point A with respect to plane α.

5. Line l passes through the point (2, 3, -4) and intersects the xy-plane at the point (5, 5, 0).

   (1) Find $a : b : c$, if the plane $ax + by + cz = d$ intersects line l orthogonally.

   (2) Find the equation of the plane in (1), if it passes through the point (4, 3, 6).

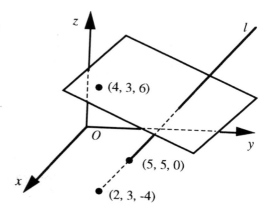

6. Determine the coefficients b, c, and d such that the plane $x + by + cz + d = 0$ includes two lines $3x = -2y = 6z$ and $2x = y = -z$ which intersect at the origin.

7. Find the equation of a plane that includes the line $\dfrac{x-1}{3} = \dfrac{y-2}{2} = \dfrac{z-1}{3}$ and passes through the point (1, -1, 2).

Answers to Chapter Exercises

Chapter 1. Quadratic Curves .... 30 – 32

A

2. $\dfrac{x^2}{6} + \dfrac{y^2}{3} = 1$

5. (1) $Q\left(\dfrac{-2x_1}{y_1 - 2}, 0\right)$

   $Q\left(\dfrac{2x_1}{y_1 + 2}, 0\right)$

B

1. (1) $(y - 3)^2 = 8x$
   (2) $(y - 2)^2 = -16(x + 1)$

2. $-\dfrac{\sqrt{5}}{2} \leq m \leq \dfrac{\sqrt{5}}{2}$

5. (1) Straight line $A'P$:
   $$y = \dfrac{y_1}{x_1 + 4}(x + 4)$$
   Straight line $P'A$:
   $$y = \dfrac{y_1}{4 - x_1}(x - 4)$$

Chapter 2. Vectors in the Plane .... 73 – 75

A

1. $P(-5, 5)$, $Q\left(-\dfrac{1}{2}, \dfrac{3}{2}\right)$

3. (1) -3 (2) -8

4. (1) $60°$ (2) $\sqrt{3}$

5. $30°$

7. (1) $\vec{p} = (2t - 6, 5t - 2)$

   (2) After $\dfrac{32}{29}$ seconds

B

1. (1) $\overrightarrow{CE} = \dfrac{1}{3}\vec{a} - \vec{b}$

   $\overrightarrow{CF} = \dfrac{1}{4}\vec{a} - \dfrac{3}{4}\vec{b}$

4. $60°$

Chapter 3. Matrices .... 120 – 121

A

1. $x = -6$, $y = 7$

2. $\begin{cases} x = 6 \\ y = -6 \end{cases}$ or $\begin{cases} x = -6 \\ y = 6 \end{cases}$

3. $k = -1$ or $k = 6$

5. $\begin{pmatrix} a & 0 \\ 0 & a \end{pmatrix}$

6. $\begin{pmatrix} -3 & -2 \\ 3 & 2 \end{pmatrix}$

7. $\begin{pmatrix} a & b \\ 2a & 2b \end{pmatrix}$
   (provided that $a \neq 0$ or $b \neq 0$)

8. (2) $\begin{pmatrix} \cos 2\theta & \sin 2\theta \\ \sin 2\theta & -\cos 2\theta \end{pmatrix}$

B

2. Straight line
   $\begin{pmatrix} x \\ y \end{pmatrix} = \begin{pmatrix} 7 \\ -2 \end{pmatrix} + t \begin{pmatrix} 5 \\ -2 \end{pmatrix}$

3. (1) x-axis: straight line $y = -\dfrac{a}{2} x$

   y-axis: straight line $y = 0$

4. $a = -2$, $b = -3$

5. $\begin{pmatrix} 1 & 0 \\ 1 & -1 \end{pmatrix}$

7. l_1 $x + y - 1 = 0$

   l_2 $3x + y - 1 = 0$

Chapter 4. Figures in Space .... 165 – 167

A

2. xy-plane: center $(3, 7, 0)$, radius 3
   yz-plane: center $(0, 7, 4)$, radius 4

3. (1) $m = 0$, $n = -9$ (2) $m = 5$

4. $2(x - 3) = \sqrt{2}(y + 1) = 2(z - 2)$

5. $2x - y - z = 2$

6. $13x - y - 7z = 24$

7. 4

8. (1) $5\sqrt{2}$ (2) $(8, 6, 10)$

B

1. $90°$

2. (1) $t = \dfrac{3(1 + \sqrt{2})}{4}$

3. $\dfrac{x + 1}{4} = \dfrac{y}{4} = \dfrac{z - 2}{3}$

4. (1) $(4, 3, -1)$
   (2) $(10, 5, -5)$

5. (1) $3: 2: 4$
   (2) $3x + 2y + 4z = 42$

6. $b = \dfrac{5}{4}$, $c = \dfrac{7}{4}$, $d = 0$

7. $11x - 3y - 9z + 4 = 0$

Index

addition theorem, 113
angle formed
    by two straight lines, 124
    by two planes, 126
asymptote, 22
axis
    of a cone, 27
    of a parabola, 3
basic vector, 43, 141
center
    of a hyperbola, 18
    of an ellipse, 11
column, 76
column vector, 77
component
    of matrix, 76
    of vector, 44, 142
component representation, 45, 143
composite mapping, 116
composite transformation, 111
coordinate, 131
coordinate axis, 131
coordinate plane, 131
difference
    of vectors, 38
    of matrices, 66
dimension, 77
directed line segment, 34
direction cosine, 145
direction vector, 153
directrix, 2
domain, 115
end point, 34
equal vectors, 35
equation
    of a hyperbola, 19
    of a parabola, 3
    of a sphere, 135
    of an ellipse, 12
focus
    of a hyperbola, 18
    of a parabola, 2
    of an ellipse, 11
force vector, 70
generatrix, 27
hyperbola, 18
identical matrices, 77
identity mapping, 118
identity transformation, 100
image, 102, 114
initial point, 34

inner product, 48, 147
intersecting planes, 126
intersection of a straight line and a plane, 127
inverse mapping, 118
inverse matrix, 92
inverse transformation, 113
inverse vector, 35, 138
line of intersection, 126
linear transformation, 101
linearity, 103
$m \times n$ matrix, 77
m rows and n columns, 77
magnitude of a vector, 35, 138
major axis, 13
mapping, 114
matrix, 77
minor axis, 13
multiplying a matrix by a real number, 81
multiplying a vector by a real number, 39
normal vector, 64, 157
numerical vector, 77
origin, 131
orthogonal vectors, 50
parabola, 2
parallel
    planes, 126
    vectors, 40
parallelism of a straight line and a plane, 127
parameter, 61, 153
parametric representation, 61, 153
perpendicular
    planes, 126
    vectors, 50
    straight lines, 124
position vector, 57, 150
position vector of a midpoint, 58
product of matrices, 84-85
quadratic curve, 27
range, 115
rectangular hyperbola, 23
rotation, 100
row, 77
row vector, 78
sphere, 135
spherical surface, 135
square matrix, 77
standard form
    of a hyperbola, 19
    of an ellipse, 12
sum
    of vectors, 37
    of matrices, 79
tangent, 8, 25

tangent point, 8, 25
theorem of three perpendiculars, 129
unit matrix, 90
unit vector, 42, 140
vector, 35, 138
vector equation, 64
velocity vector, 71
vertex
    of a cone, 27
    of a hyperbola, 20
    of a parabola, 3
    of an ellipse, 13
zero factor, 90
zero matrix, 80
zero vector, 38, 138

Table of Squares, Square Roots, and Reciprocals

n	n^2	$\sqrt{n}$	$\sqrt{10n}$	$\frac{1}{n}$	n	n^2	$\sqrt{n}$	$\sqrt{10n}$	$\frac{1}{n}$
1	1	1.0000	3.1623	1.0000	51	2601	7.1414	22.5832	0.0196
2	4	1.4142	4.4721	0.5000	52	2704	7.2111	22.8035	0.0192
3	9	1.7321	5.4772	0.3333	53	2809	7.2801	23.0217	0.0189
4	16	2.0000	6.3246	0.2500	54	2916	7.3485	23.2379	0.0185
5	25	2.2361	7.0711	0.2000	55	3025	7.4162	23.4521	0.0182
6	36	2.4495	7.7460	0.1667	56	3136	7.4833	23.6643	0.0179
7	49	2.6458	8.3666	0.1429	57	3249	7.5498	23.8747	0.0175
8	64	2.8284	8.9443	0.1250	58	3364	7.6158	24.0832	0.0172
9	81	3.0000	9.4868	0.1111	59	3481	7.6811	24.2899	0.0169
10	100	3.1623	10.0000	0.1000	60	3600	7.7460	24.4949	0.0167
11	121	3.3166	10.4881	0.0909	61	3721	7.8102	24.6982	0.0164
12	144	3.4641	10.9545	0.0833	62	3844	7.8740	24.8998	0.0161
13	169	3.6056	11.4018	0.0769	63	3969	7.9373	25.0998	0.0159
14	196	3.7417	11.8322	0.0714	64	4096	8.0000	25.2982	0.0156
15	225	3.8730	12.2474	0.0667	65	4225	8.0623	25.4951	0.0154
16	256	4.0000	12.6491	0.0625	66	4356	8.1240	25.6905	0.0152
17	289	4.1231	13.0384	0.0588	67	4489	8.1854	25.8844	0.0149
18	324	4.2426	13.4164	0.0556	68	4624	8.2462	26.0768	0.0147
19	361	4.3589	13.7840	0.0526	69	4761	8.3066	26.2679	0.0145
20	400	4.4721	14.1421	0.0500	70	4900	8.3666	26.4575	0.0143
21	441	4.5826	14.4914	0.0476	71	5041	8.4261	26.6458	0.0141
22	484	4.6904	14.8324	0.0455	72	5184	8.4853	26.8328	0.0139
23	529	4.7958	15.1658	0.0435	73	5329	8.5440	27.0185	0.0137
24	576	4.8990	15.4919	0.0417	74	5476	8.6023	27.2029	0.0135
25	625	5.0000	15.8114	0.0400	75	5625	8.6603	27.3861	0.0133
26	676	5.0990	16.1245	0.0385	76	5776	8.7178	27.5681	0.0132
27	729	5.1962	16.4317	0.0370	77	5929	8.7750	27.7489	0.0130
28	784	5.2915	16.7332	0.0357	78	6084	8.8318	27.9285	0.0128
29	841	5.3852	17.0294	0.0345	79	6241	8.8882	28.1069	0.0127
30	900	5.4772	17.3205	0.0333	80	6400	8.9443	28.2843	0.0125
31	961	5.5678	17.6068	0.0323	81	6561	9.0000	28.4605	0.0123
32	1024	5.6569	17.8885	0.0313	82	6724	9.0554	28.6356	0.0122
33	1089	5.7446	18.1659	0.0303	83	6889	9.1104	28.8097	0.0120
34	1156	5.8310	18.4391	0.0294	84	7056	9.1652	28.9828	0.0119
35	1225	5.9161	18.7083	0.0286	85	7225	9.2195	29.1548	0.0118
36	1296	6.0000	18.9737	0.0278	86	7396	9.2736	29.3258	0.0116
37	1369	6.0828	19.2354	0.0270	87	7569	9.3274	29.4958	0.0115
38	1444	6.1644	19.4936	0.0263	88	7744	9.3808	29.6648	0.0114
39	1521	6.2450	19.7484	0.0256	89	7921	9.4340	29.8329	0.0112
40	1600	6.3246	20.0000	0.0250	90	8100	9.4868	30.0000	0.0111
41	1681	6.4031	20.2485	0.0244	91	8281	9.5394	30.1662	0.0110
42	1764	6.4807	20.4939	0.0238	92	8464	9.5917	30.3315	0.0109
43	1849	6.5574	20.7364	0.0233	93	8649	9.6437	30.4959	0.0108
44	1936	6.6332	20.9762	0.0227	94	8836	9.6954	30.6594	0.0106
45	2025	6.7082	21.2132	0.0222	95	9025	9.7468	30.8221	0.0105
46	2116	6.7823	21.4476	0.0217	96	9216	9.7980	30.9839	0.0104
47	2209	6.8557	21.6795	0.0213	97	9409	9.8489	31.1448	0.0103
48	2304	6.9282	21.9089	0.0208	98	9604	9.8995	31.3050	0.0102
49	2401	7.0000	22.1359	0.0204	99	9801	9.9499	31.4643	0.0101
50	2500	7.0711	22.3607	0.0200	100	10000	10.0000	31.6228	0.0100

Greek Letters

Capital	Small		Capital	Small	
A	α	alpha	N	ν	nu
B	β	beta	Ξ	ξ	xi
Γ	γ	gamma	O	o	omicron
Δ	δ	delta	Π	π	pi
E	ε	epsilon	P	ρ	rho
Z	ζ	zeta	Σ	σ, ς	sigma
H	η	eta	T	τ	tau
Θ	θ, ϑ	theta	Y	υ	upsilon
I	ι	iota	Φ	φ, ϕ	phi
K	κ	kappa	X	χ	chi
Λ	λ	lambda	Ψ	ψ	psi
M	μ	mu	Ω	ω	omega